MEDIA for INTERACTIVE COMMUNICATION

MEDIA for INTERACTIVE COMMUNICATION

Rudy Bretz

with

Michael Schmidbauer

SAGE PUBLICATIONS
Beverly Hills / London / New Delhi

For information address:

SAGE Publications, Inc.
275 South Beverly Drive
Beverly Hills, California 90212

SAGE Publications India Pvt. Ltd.
C-236 Defence Colony
New Delhi 110 024, India

SAGE Publications Ltd
28 Banner Street
London EC1Y 8QE, England

Printed in the United States of America

Library of Congress Cataloging in Publication Data

Bretz, Rudy.
Media for interactive communication.

Bibliography: p.
1. Teleconferencing. I. Schmidbauer, Michael.
II. Title.
TK5102.5.B693 1983 384.55'564 83-3085
ISBN 0-8039-1986-7

FIRST PRINTING

CONTENTS

Acknowledgments 7

Introduction 9

PART I: SOME FAMILIAR INTERACTIVE MEDIA SYSTEMS 29

1 Television Systems 33

2 Audio Systems 69

3 Augmented Audio Systems 80

4 Alphameric Media Systems 93

5 Data Response 110

6 Slow Scan TV 117

PART II: QUASI-INTERACTIVE MEDIA SYSTEMS 137

7 Teletext and Reuters IDR 143

8 Videotex 154

PART III: SYSTEM DESIGN FEATURES: ISSUES AND PROBLEMS 167

9 Technique: The Use of the Means 169

10 Intrinsic Factors in Teleconference System Design 174

11 Extrinsic Factors in Teleconference System Design 220

12 What Medium for What Purpose? 231

13 Developments to Come 240

Glossary 247

References 255

Index 259

About the Authors 264

ACKNOWLEDGMENTS

As senior author, I would like to thank Michael Schmidbauer and the Institut für Telekommunikation, Munich, West Germany, for instigating the work that led to this publication. Without the support of the Institut für Telekommunikation, under Dr. Schmidbauer's inspiring guidance, this work could not have been accomplished. I would also like to thank my colleague Dr. Robert Anderson, who had the interest and took the time to read the entire book thoroughly and contributed many excellent suggestions for improvement. Lawrence Lipsitz, publisher of *Educational Technology,* generously allowed the chapter on slow scan TV to be produced from its original publication in the above magazine. I want to thank Glen Southworth of Colorado Video, the outstanding pioneer in the slow scan TV field, for checking the chapter on slow scan TV and preparing a set of illustrations for it.

— *Rudy Bretz*

INTRODUCTION

Most communication among persons is interactive. Put two or three people together, introduce them, and interactive communication begins. People together, in nearly any culture, will, if you let them, converse more than they will do anything else. Children on the street, families in the home, workers on the job are constantly exercising the power of speech and the sense of hearing. The sense of sight, no less active during conversation, is clearly less important. A deaf person without lip-reading ability or sign language is far more isolated from the world of people than a person who is blind. Customs vary, but in most western societies about the only set of circumstances that will prevent conversation between two normal peole who speak the same language and are within conversing distance of each other is that the place is public, the people have not been properly introduced, and both of them are too reserved to overcome these barriers. In a private or semiprivate place, such as a doctor's office or a path in the woods, shyness or not knowing each other is easily overcome by the conversational imperative.

But there are other communication media that compete. If a person spends most of his or her leisure hours in interactive conversation, the chances are that he or she spends the rest in passive reception of mass media: reading books or magazines, watching films, television programs, shows of other kinds, listening to the radio or to music. This need for entertainment, for stimulation, for vicarious experience is great enough to have spawned great mass communication industries: Broadways, Hollywoods, national radio and television networks, press and publication empires. Yet, compared with the need for personal interaction, this need is relatively small. People can live without one-way mass media if they have each other, but total solitary confinement for a long enough time can drive a person mad even if he or she has a television set to watch, although it may take longer in that case.

Mass communication has practically always been one-way. "Two-way mass communication" would be considered a conflict in terms. The closest we have come until recently is the radio/telephone program. When President Carter, in the early months of his administration, held radio and television interviews for the public, anyone who wished could talk to the president. Of course, only a few of the thousands who tried could get through and the listening millions were still just that, passive receivers of a one-way program.

One-way communication did not originate with the mass media. Modern technology has merely extended the entertainment and information that once was presented directly in front of audiences around campfires, in village meetings, at court, and in theatres, and has brought it to everyone in his or her home. But what has technology done for *interaction* — the one-on-one and the small group situation?

The telephone is the only outstanding example. It took nearly a hundred years for the telephone to become ubiquitous, even in the western world, but now it is a necessity for the homemaker as well as for the business person. And despite its availability to the great mass of people, nobody thinks of the telephone as a mass medium.

The mass media are few; newspapers, magazines, radio, and television just about complete the list. The telephone is used in a few places for one-way information transmission, but this amounts to no more than a fraction of 1 percent of its use — a very small fraction at that. The telephone is a two-way interactive medium.

Technically, radio is very similar, but in use altogether different. A comparison of these two audio media, familiar to us all, will help explain just what interactive communication is and what it is not.

Technically, radio and telephone are basically the same. They both use microphones and loudspeakers, although those built into the telephone handset are very small and of limited quality in comparison with those used in radio. The transmission systems of the two are likewise similar. While radio implies broadcast over the air waves, it is also distributed by wire from central studios to broadcast stations and between cities. While telephone implies wire and cable distribution, much of the telephone traffic is now carried by microwave beams along land routes, or over air-wave routes thousands of miles long to and from communication satellites.

The most significant technical difference between radio and telephone is that radio is distributed in one direction only, and thus requires only one wire or broadcast channel, while telephone requires two, operating in opposite directions. (Some inexpensive telephone systems use only one channel and users must "push to talk," thus reversing the direction of the system between an utterance and reply, but this is far less satisfactory and used only where necessary.)

Another salient difference between radio and telephone is that the content of radio is generally a preplanned and prearranged "program." For convenience this is often prerecorded; almost always it is scheduled. Some programs are prepared by writers and produced by professional "talent" of one kind or another, so that every word that is uttered is first committed to paper. Other programs are only outlined, thought out, and paced, but the actual spoken words are extemporaneous. Still, just about all the content of radio consists of programs, while just about all the content of telephone communication does not. There are exceptions, of course. There is no technical reason that a two-way medium cannot be used for one-way presentation of programs, and since the telephone distribution system is already in place in nearly all homes, it forms a very convenient means to distribute such things as health information programs, instruction, prayer and inspirational messages, and weather and time data. For example, the

London telephone directory contains a list of numbers to call for one-way services, from bedtime stories to information about what's going on in town.

In summary, we can say that radio is a typical mass medium. The same program content, or "software," is distributed to a large number of people. It is a one-way, few-to-many communication system. The telephone, by contrast, is a one-to-one communication medium in which messages are intended for and (barring wiretapping) reach only one party. It is not considered a mass medium despite the fact that nearly as many homes have telephones as have radios in the United States, Canada, and European countries.

To fully understand interactive communication we must continue this analysis a little further. In a one-way mass medium, such as radio, one person (or group of people) plays the role of sender, while a larger number of people play the role of receiver. These roles are not interchangeable — except for the case of the listener who participates in a telephone talk program. In an interactive situation, on the other hand, the sender and receiver roles and interchanged with each utterance.

The term "interactive" is sometimes used for systems in which these roles are not totally interchangeable. In some people's views, interaction requires two actions: one by the sender conveying a message (statement, command, question, or whatever) and the other by the receiver in responding to the message. Thus some teaching machines that present a frame of information, ask a question, and refuse to proceed until the learner selects the correct response are called interactive by their manufacturers. I prefer to refer to such systems as "active" or "responsive," or, by the more general term, "quasi-interactive." They may give a strong illusion of being interactive but they are only "quasi" so.

Another example is the system in which the two actions are (1) the selection (by the user) of a particular program or piece of information he or she wants to receive, and (2) the provision (by the sender or some automated sending device) of the desired material. Information-retrieval systems that respond to requests keyed in by users are an example of this type of so-called interactive media.

In our view, two actions are not enough to characterize a system as interactive; there must be three: (1) a statement, question, request, or other message from communicant A to communicant B, (2) a response from B to A based on what A has just said, and (3) a response or reaction from A to B based on B's response. Of these three actions, the second and third are always required in interaction; the first is dispensable. Only at the beginning of an interchange is the first action necessary, or at the start of a new subject of discussion. The rest of the time the two communicants respond *to each other.* This is the best definition of interactive communication that we have been able to come up with, and it is the one we shall use in this book. *Interactive communication is the situation in which each of two (or more) communicants responds to the other.*

Here are some further examples of active or responsive communication that, according to this definition, are not interactive.

In a television "sing along" program, the words of a song are shown on the screen and the viewers are encouraged to sing along. This is the active or responsive mode of quasi-interaction. The first two actions are present, but there is no third action, except in illusion, as when the "conductor" encourages the viewers by telling them they are doing well. He or she actually has no knowledge of whether they are doing well or not.

The branching story, pioneered on the stage and in films notably by the Czechoslovakians, is an example of the selection mode of quasi-interaction. The story is presented up to a dramatic moment when more than one course of action is possible. The film (or stage action) is stopped and the audience votes on which of two possible turns they would like to see the plot take. A second projector and film (or an alternate rehearsed script) are ready in case an alternate course is voted for, and the two routes join again in a later sequence, or, like a tree, continue to branch apart at each decision point (a far more expensive production task).

Improvisational performances in which actors are given a situation and create their own roles and dialogue on the spot have involved theatre audiences in selective quasi-interaction for years. Characters, situations, and action are suggested by members of the audience. To some extent these entertainments have been adapted for one-way television, but without the participation of the home audience.

A children's television series of an earlier decade, called *Winkie*, demonstrated a very ingenious response idea. To participate, a child had to acquire, usually by mail, a clear plastic sheet and crayon. The sheet was attached to the television screen by smoothing it on, and was held there by static electricity. Then the child took part by drawing key parts of the picture. For instance, when the cartoon character Winkie was caught in a corner of the castle where there was no escape, and the villain was rapidly closing in, the announcer would say, "Quick, boys and girls, draw a door in the wall on the left, quick, a door so Winkie can escape!" After a reasonable few moments for some frantic drawing, and urging, the predrawn and prerecorded Winkie opened the door and escaped. Then the children were urged to quickly rub out the door they had created so Winkie would not be followed, and were delighted to see the villain baffled.

Now for an example of true interaction.

The old-time face-to-face storyteller, represented in modern society by the father telling a bedtime story to his small children, may find himself interrupted, either by corrections — if the children have heard the story before in a slightly different version — or, if it is a made-up story, by suggestions and requests ("Daddy, make it that the dog catches the bear and eats him all up!"). He may even have to respond to questions about what happened next if the story seems to wind up open ended. Only in this last example of interactive entertainment do we have a case of true interaction, with each party responding to the messages of the other. After all, this is a face-to-face situation, in which there is always the potential for true interaction.

Simulated Interaction

It does not make the definition too broad to include under the term "communicant" nonhuman as well as human intelligence. Intelligence there must be, of some order, if messages are to be perceived and comprehended and responses specific to these messages are to be conceived, formulated, and transmitted. Such intelligence may be found in humans, on a wide range of intelligence levels, or in sophisticated machines that have been so elaborately programmed that they seem human. The machine, still far simpler than the human, has only a cognitive intelligence — only a left brain, if you will. The right side of the buman brain, which handles affective matters, art and emotion, and is usually the source of the power that moves us both mentally and physically, is also frequently guilty of disrupting the logic of the left brain's cognitive thinking. With no right brain, the computer avoids such disruption. Barring short circuits and downtime, the computer already far outdistances the capability of the left human brain in matters of calculation and logical thinking. As an intelligent tool, the computer can be provided with considerable communication capability, and hence it is indeed fully possible for a human to interact with a computer under conditions in which each may respond and react to the other.

One particularly well-written computer program, called ELIZA, simulated a psychotherapist drawing out the feelings of a patient. The computer user, playing the role of patient, answered questions posed by the program. The computer was programmed to select key phrases from each reply made by the patient and incorporate these into further questions, such as: "How did you feel about that?" and "Why do you think you felt that way?" The result was so realistic that many people, doctors specifically, received the strong illusion that they were interacting with a real person; some refused to believe that their interlocutor was only an intelligent machine. Such simulated interaction is also used in business and industry. Some savings banks, for instance, operate 24 hours a day, making transactions at night with the aid of appropriately programmed computers.

Some Necessary Definitions

The next few pages are necessary to assure successful one-way communication between author and reader during the remainder of this book. One person cannot understand verbal communication from another unless he or she puts the intended meaning on at least the key words. The study of semantics has shown us how often a word will have different meanings for different people. In many cases dictionaries will recognize the most common of these and list them as separate senses.

"Message"

Some people use the term "message," for example, in its broadest sense, to mean anything that is communicated. They will even call entire programs messages, using the term as a synonym for content. This is the first sense listed in a typical dictionary definition, for example, "1. A communication transmitted from one person or group to another" (American Heritage Dictionary, 1970). Other people take a narrower meaning. They contend that "message" implies meaning, in terms of information, ideas, or feelings. It is possible, even frequent, they say, for programs to have such a simple entertainment value that they have no informational or idea content at all beyond what is commonplace to all receivers, and hence cannot be called messages. A message, in this view, could be something contained in a program, but not the program itself. A meaningless program would thus carry no message.

A third meaning, narrowest of all, defines "message" as "the basic theme or significance of something" (American Heritage Dictionary, 1970). This, we would guess, is the sense in which Marshall McLuhan (1964) used the word in his famous phrase: "The medium is the message." Because this phrase is so famous, a short discussion of it is in order. To those to whom "message" meant the content of a communication, the phrase was either meaningless or interpreted as cryptic on the generous basis that the man must have had something in mind — it was "for him to know and for us to find out." To those also to whom "message" meant the meaning contained in a communication, the phrase was baffling. McLuhan, I think, actually coined his own subdefinition for the term. He meant it to mean theme or significance; not the significance of a communication (letter, program, speech, statement, or the like), but the social significance of the medium itself ("medium" being all the technology for the production, transmission, and reception of communications, and all the people involved). He was pointing out that the pure existence of a medium is in itself more important than any use that is made of it. This is probably true.

For instance, Henry David Thoreau and Ralph Waldo Emerson, great American men of letters at the time when the first telegraph line was built from Maine to Texas, are today both given credit for the sage remark: "Perhaps Maine has nothing to say to Texas." Whoever said this did not understand, or anticipate, that the telegraph, and the other telecommunication media that followed, would make possible a new level of communication that would transform western society. As soon as the means of transmission were available, Maine thought up lots of things to say to Texas. No one but McLuhan, however, would have thought of calling this significance the "message," and in this book we will not follow his lead.

The content of interactive communication is hardly a "program," in the sense of a prepared presentation, so we can eliminate that source of confusion; we will not be using "message" to mean "program." In interactive communication, someone is usually talking with someone, showing something, asking questions, giving information, giving commands, and the like. The purpose of interaction is not entertainment in the usual passive sense, however entertaining conversation may be. We have to assume that for every spoken utterance, whether face-to-face with others or via some interactive medium, there is a purpose in mind and some meaningful content. Thus for the purposes of this book we can equate "message" with content and say the communication is the message, whatever that communication may be.

There has been *responsive* communication when a receiver begins a dialogue with the sender, reversing roles and sending a message back to him or her. A receiver could respond in other ways, with thoughts or actions, but this does not count as communication unless he or she sends a message. When this response message is received and *reacted to by the original sender,* and this reaction to the response makes the full journey again, finally being received and comprehended by the original receiver, then and only then can we say there has been fully *interactive* communication.

But there is still another question we must ask before we close this subject. Can there be interaction that is not communication? Of course there can. When two children throw a ball back and forth, or when they fence, box, or wrestle, they are certainly interacting. Electronic games are interactive without being used for communication. When I touch a button and cause a distant elevator to rise to my floor, or when a NASA sends a signal to a distant space probe to change its course, there is interaction. There is however, no cognitive message (beyond the simple information that the command has been received and acted upon) and hence no communication in the sense in which we are using the term.

Engineers would disagree. "Communication theory" is a mathematical concept originated by engineers, and revolutionary in its usefulness in engineering, but it remains almost exclusively in their province. In communication theory the term "message" is used to mean anything at all transmitted from one place to another (or recorded). Any transmission of anything except "noise," even a single bit (which can be signaled by a change in the voltage level of an electrical current) can be called information and considered a message. Note, however, that only *transmitted* information (what we are calling *tele*communication) qualifies as communication in the engineering sense; recorded information that is not subsequently transmitted (such as written or printed materials, audio- and videotapes, and the like) is relegated to another category.

It is obvious that the engineer on the one hand and the philosopher (like McLuhan) on the other use these terms quite differently from the way this book will use them, and it is important for us to get these signals straight before we begin.

"Communication," "Communications"

It is probably unnecessary to discuss the verb "communicate" because there is only one appropriate dictionary definition and this is wholly acceptable for this book. Such descriptors as "make known," "impart," "transmit," and so on express the same idea. But some of the derivatives of "communicate" should be discussed.

"Communication" (noun) is, of course, the act of communication — "the exchange of thoughts, messages, or the like, as by speech signals or writing," according to the 1976 American Heritage Dictionary. A second sense, in which we will also use the word, is "that which is communicated; a letter or message" (Funk and Wagnalls Dictionary, 1970). But what of the word "communications"? This word can mean "several messages," or it can mean much more than that, a sense in which it is frequently found in technical books. One dictionary defines this word as "means of communicating," which is a rough description of the technical side of communication, exclusive of the content. "Communications" is often used to mean only the electronic means of transmission, particularly by engineers and technical people. When used in this book it will mean only the plural of "communication," in the sense of more than one message.

"Medium"

Another key word that should be defined is "medium," for which some dictionaries list up to thirteen senses, several of which are appropriate to communication. Here again it is important that both author and reader understand the same meanings for the term, especially when it is to be used in two or more senses.

The idea of intervention and intermediacy appears often in dictionary definitions of "medium"; the word "means" is frequently used. What I intend to mean by "medium" is more specific than these definitions, but would be covered by almost any one of them.

For the purpose of this book, a medium is a technological system for conveying messages, operating intermediately between sender(s) and receiver(s), when they are separated in space, time, or both. This includes means for picking up the message, recording and transmitting it, and displaying it (usually) for the eye or for the ear of its intended recipient(s). It includes in most cases hardware (equipment), software (film, tape, disc, paper, and so on), and the creative and technical people involved in operation and production (sometimes called "peopleware").

"Teleconference"

Two terms nearly as familiar as "interactive communication" itself are "teleconference" and its derivative "teleconferencing." These have acquired several meanings as well, although all are closely related. It seems that the practitioners of each interactive medium have tended to adopt the term as their own. At the University of Wisconsin, where the existing telephone network was used for interconnecting the most extensive interactive network in the world, the prefix "tele" is claimed as an abbreviation for "telephone." Some practitioners of video systems claim it is an abbreviation for "television." Computer communicators use "teleconference" to refer to their alphameric medium (akin to electronic mail), with which isolated individuals at their own terminals may interact in a far-flung group by typing and print display, either in real time or by entering and recording messages in jointly used computer files. In fact, the term is really generic, not specific to any medium. Prefixed by the Greek word for distance, it applies to all media that transmit across space, in real time.

This prefix has acquired a second sense, however; "tele" has come to mean "electronic." This is the sense in which computer people use the term. To many people today "telecommunication" means electronic communication. They would not include the eighteenth-century visual semaphore, for example, even though a semaphore relay system could send messages over long distances in short order. Computer conferencing that consists entirely of recorded messages, recording at one time and reading later, is electronic and hence included even though it does not operate in real time. In this book both these senses will be accepted, and by context the reader should have no difficulty understanding which is meant.

The reader will encounter many other terms in this book that are peculiar to interactive communication or communication in general. Each will be defined when first used. A glossary of the least familiar terms used in this book is also included at the back of the volume.

Media for Interactive Communication

Recording (storage) media, such as writing and print in their various forms, or audiotape recording, have not been too satisfactory for interactive communication. The real value of recording something lies in its access for future reference, either at the same place or elsewhere after the recording has been transported. In any case, time will intervene between the sending of a message and its reception, and further time elapses before a responding message can reach the original sender — by which time he or she often has forgotten what was said to begin with.

Interaction generally works best when there is the least time lapse between the sending of a message and the reception of its reply. Except for the cost, most people would probably use telephone calls for personal correspondence instead of mail. Indeed, the long-distance phone is heavily used by firms that engage the low-cost WATS (wide area telephone service) lines; the more cumbersome and time-consuming letter correspondence medium is used mainly as an adjunct to the phone, when matters of agreement and confirmation need to be put down on paper.

Truly interactive mail correspondence is rare, even between good correspondents. Because of the time lag, writers must keep copies of their letters or rely on their correspondents to repeat questions before answering. Probably most personal letters are designed to stand alone and hence are more like one-way presentation than two-way interaction. Business letters are rarely written unless there is something that needs to be recorded; the telephone is used instead. There is something more formal and elegant about the written (or printed) word, however; good etiquette gives preference to thank-you notes and written invitations over telephone calls.

In interaction, messages must flow back and forth, to and from specific individuals or small groups of people. In natural face-to-face relations this is to some extent simultaneous; a listener shows constant response through facial expression and one-syllable vocal utterances. Reply can be immediate, and usually is. To be useful for interactive conversation, media must have at least the capacity for immediate response, if not simultaneous response.

An interactive system always includes, at the least, two channels facing in opposite directions, or a single channel that is reversible. The telephone, on which you can both talk and listen at the same time, is an example of the first. The second is exemplified by the push-to-talk intercom system, on which you can either listen or talk, but not both at once. A cheap "squawk box" will normally be in the receive mode ready to display (in sound) whatever messages may be sent, but when you push the "talk" switch you turn the small loudspeaker into a low-quality microphone and transmit back down the same wire.

Interactive systems are generally closed-circuit, meaning that they are not broadcast to the general public, although broadcast channels *may* be used for all or part of an interactive system. The telephone, again, is an essentially private wired system, with elaborate automated central switching apparatus to keep large numbers of private binodal circuits (each consisting of two channels) separate from each other.

Citizen band radio, which gets more and more public as its popularity increases, devotes a few broadcast channels within a given area for all users. The classic example of interactive radio is the Flying Doctor Service of central Australia. Every isolated sheep station has its transceiver. Like a multilocation intercom system, the various regions of the Australian outback are interconnected in one great party line for medical assistance, education, and just plain sociability. People who live a hundred miles apart know each other and each other's business extremely well in the outback — much better, it might be said, than next-door neighbors in most urban areas.

Using a medium interactively imbues it with some of the salient characteristics of face-to-face conversation. With push-to-talk systems immediate response is possible, but not while the first person is still speaking. With two-way systems such as the telephone, the simultaneity of transmission in both directions adds some important factors. Simultaneous feedback of grunts and short utterances from a listener tell the speaker he or she is still there and attending, and gives some clue to the listener's reactions. There is the normal conversational capability for the listener to interrupt a speaker. In order to save time the listener may demonstrate that he or she understands by providing an example; he or she may stop the speaker on a point that needs expansion. Finally, there is the capability for the normal overlap of voices that occurs in conversation, as when a listener starts to reply before the speaker gets his or her last words out. Sometimes in an animated face-to-face conversation two participants will make statements at the same time, and each will hear and understand the other. This freedom, possible on the telephone, must be inhibited when using a push-to-talk system, and to that degree such systems are unnatural. As we shall see in later discussions in this book, the advantages of simultaneous feedback do not particularly increase the amount of information that can be carried on a system, but they do make it more natural, allowing people to talk as they do when face to face and hence making the system itself less noticeable to the user. This is sometimes called "transparency" of a system.

One other characteristic of interactive communication should be mentioned here, to further distinguish it from one-way communication. The content of interactive communication tends to be informal, usually called "ad-lib," which means it is free of previous planning. The content of one-way media, on the other hand, is based on the concept of the "program," a presentation to an audience, usually planned carefully in advance and usually formal in nature. Words are generally scripted in advance and then either memorized or read. Even informal radio disk jockeys present programs that consist of prerecorded music, the order and pacing of which represents much prior thought and effort.

Media Versus Means

Communication *means* are the ways we have of forming messages; it is impossible to communicate without using some *means*. Communication *media*, on the other hand, may be used to record or transmit such means, in order to bridge time or space, but face-to-face communication can take place without media.

In face-to-face communication, when communicants share the same time and space and can see and hear each other, they use a variety of natural and artificial means of communication. By "natural" means, I mean those that a person is born with. Natural means of communication originated far back in our animal ancestry and are not peculiar to human beings. Facial expression, body posture, actions, nonverbal utterances, and pointing are examples of natural means. Most of the artificial means of communication originated with primitive humans and include language, drawing and painting, music, and dancing. These are all culture related and have to be taught. I call these, both natural and acquired means, *basic* means of communication. Since language is as important as all the rest put together, they are often classified into two categories: the verbal and the nonverbal.

None of these means of communication are media. They are more basic than that, basic to media and basic to face-to-face (nonmedia) communication. They are means of formulating messages — means of expression. Media, by contrast, are technological systems for transmitting and/or recording such means. If a message is expressed in one or more of the basic means, spoken words, facial expression, and gestures, for instance, it may be transmitted by media such as television, or recorded on tape or film, and thus reach people who are not present in the same place and times (see Figure 1). The spoken words alone may be transmitted by radio or telephone and/or recorded on audio tape or disc without the facial expressions, gestures, and so on that accompanied them. If the message is a simple transfer of information, this may not be a serious loss. If the face and hands of the speaker are not being used for communication, they can actually be a distraction. (There is considerable experimental evidence that tasks involving verbal information only, such as questioning, problem solving, and the like, are more readily accomplished using audio alone than using audiovisual systems, such as television.)

One of the earliest media, and so far the most important — written language — can be used to further illustrate this point. To use this medium, messages were first conceived in spoken language, an artificial basic means of communication. If the writer did not actually speak, he or she certainly had to covertly express a thought in speech before anything could be written down. Speech, a basic *means* of communication, was thus incorporated into the technological *medium* of writing, which developed over the years into movable type, alphameric keyboards, linotype,

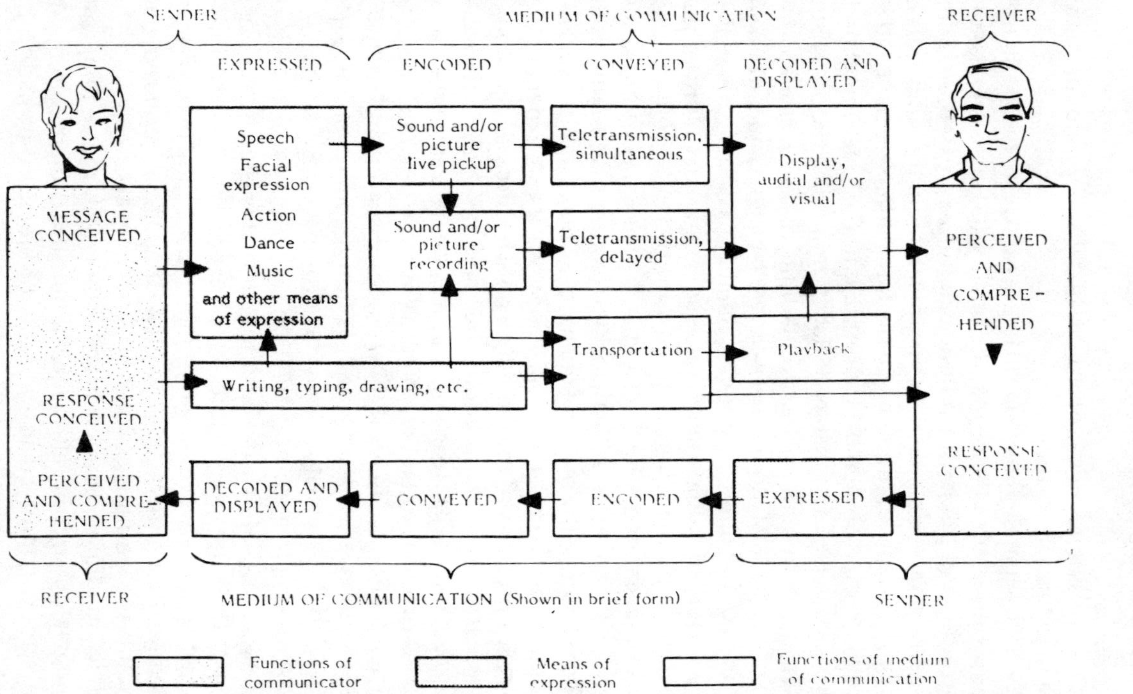

Figure 1: A media system model that distinguishes between means of expression and media of communication.

photo composition, digital recording, text displayed on the television screen, and so forth.

When we separate the means of expression from the medium in our thinking, we see at once that the quality and effectiveness of a message depend primarily on how well it is expressed, and only secondarily on the characteristics of the medium that transmits or records it. Of course, different media have different capabilities. The medium used should be appropriate to the means of expression (or vice versa). For instance, dance requires a motion-visual medium, music requires an audio medium, and so forth.

Failure to understand this very simple concept of means versus medium once led to the channeling of millions of research dollars into comparative studies (media versus face-to-face) for the purpose of determining whether television (or film) could "teach." This was a misguided effort, since it is the basic means of expression and the responsive activities of the learner that do the teaching (and learning); the medium mainly distributes and delivers the teacher and the teaching materials. For years it was expected that some magic property of the medium would be discovered through such research, something about the medium itself that would somehow improve the teaching that was being transmitted — as measured by the resultant learning. When the outcome was "no significant difference" in nearly every case, observers and sponsors of the research were disappointed, as though the researchers had failed to find the touchstone that was promised.

Most interactive communication, whether face-to-face or mediated, takes place via the basic means that existed in the Stone Age: spoken language, gesture, facial expression, and so forth. One-way media, however, often require the learning of new skills. The first recording medium — written language — required, and still requires, a learning period measured in years. Telegraph, the first "tele" medium, required yet a further skill, the sending and receiving of Morse code, although learning the code required only a minor effort compared to the skills of literacy that had to be acquired first. Learning to speak into a microphone or camera is hardly comparable to learning to read and write, yet takes hours of practice.

If the use of a medium by a teacher or other presenter or performer requires new skills, it is not because a medium is being used, but because it is a *one-way* medium. Speaking to people that you cannot see, who may or may not be listening or will not be listening until sometime in the future, requires powers of imagination not needed when you are talking to someone in front of you. You need to imagine and visualize some target audience, either an individual or a group, or your speech will not seem natural either to you or to others. Give a person a tape recorder and microphone for the first time and he or she will be nonplussed. He or she will not know how to communicate in a one-way mode. Despite years of verbal, face-to-face communication, he or she will become tongue-tied. If he or she does attempt to say something, the chances are that the words and phrases chosen will be quite different from those used in ordinary conversation, more formal and stilted, as though the role of radio announcer is called for. The same stiffness characterizes most people when they face a radio microphone or television camera for the first time. One-way speech, for the untrained, is so uncomfortable that many people refuse to leave recorded messages when they call a friend and are greeted by a telephone-answering machine.

A good audio-based telecommunication system, on the other hand, will seem perfectly natural to people who use it for the first time. No new skills are needed in audio-based systems and the user relates easily to the person with whom he or she is conversing. The intervening media system can be so transparent that the user is largely or wholly unconscious of it. Some early demonstrations of the telephone made much of this naturalness. In one case a pair of Choctaw Indians were presented conversing via telephone in their native language, dramatizing how natural it was to use the new device, even for people with no special education or training.

Various media systems and configurations of equipment differ in how readily their users forget and ignore the presence of cameras and microphones, for instance, seeing *through* the system to concentrate wholly on the ultimate receiver to whom they are communicating. This quality is called the "transparency" of a medium, and is a characteristic of a specific system configuration, not of a medium in general. Thus an audio system in which microphones are remotely located, hidden, or fastened to the user's chest will probably be more transparent than systems in which a large microphone is placed directly in front of the speaker, or in which the speaker is required to hold the microphone to his or her mouth when speaking.

All communication media in use today are limited to carrying stimuli to the eye, the ear, or both. Tactile sensation, for instance, cannot be included, except by implication. There is still no telecommunication for the handshake. (However, the "telehandshake" is not impossible by any means. All the necessary technology is already in existence; no technical breakthroughs are needed. A possible configuration of such a device is described in Chapter 13.) There are other natural means of nonvisual, nonverbal communication between people that we are only beginning to study and that we do not yet understand. It seems likely, however, that most of these minor and extrasensory means are primarily important to the affective components of a message, and to attitudes about the message sender and message receiver, rather than to the cold information content of the transmission. So far as mere cognitive information is concerned, audial and visual means, properly used, appear adequate for all our present communication needs.

The essential content of this book is to be found in Part III, in the section on system design features, the issues and problems that characterize these features, the analysis of causes for success and failure, and design suggestions for future systems. To make these concepts and comparisons understandable it is necessary to first describe a few representative examples of the various kinds of interactive systems that have been used, starting with the most familiar types, such as video and audio conferencing, and progressing to minor media systems that are less well known.

Part II is an extended section on what we call "quasi-interactive" media systems. These relatively new systems bring information in alphameric and graphic form into the home or office, transmitted in between the video frames on standard broadcast channels or fed via regular telephone circuits. Since these systems are not at present used for teleconferencing (although there is no reason they could not be used in full interactive mode in the future), this part of the book could well stand alone and need not be considered as a necessary introduction to Part III, System Design Features, where the real meat lies.

SOME FAMILIAR INTERACTIVE MEDIA SYSTEMS

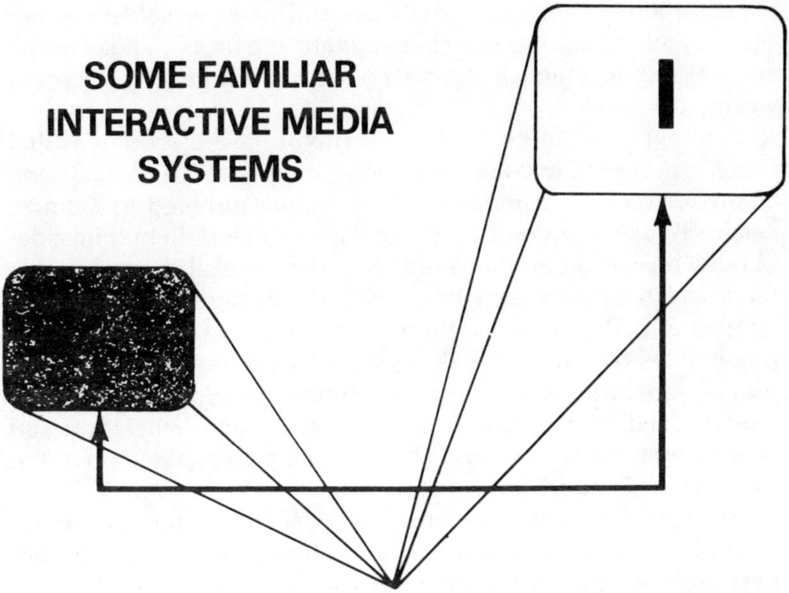

The interactive media systems described in this section are all based on words, either spoken or written. This is because almost the entire content of interactive communication is cognitive, and for cognitive information transmission words are the most important means. Thus for our taxonomy the interactive media will be divided between those that are based on (1) the spoken word, for which the technical medium of transmission or recording is audio, and (2) the written word, the general medium for which we will call alphameric.

These two categories are not mutually exclusive; printed words are an important component of most audiovisual presentations. The key phrase in the taxonomy is "based on." Typical audiovisual presentations, even those that contain pictures and diagrams, rely on the spoken word to put across the cognitive concepts involved. Lose the sound track and you have lost all. Lose whatever lettering is on the screen and you have lost only an aid — an enrichment or supplementary factor; you can generally follow most of the exposition from the sound track alone if you listen carefully.

A medium that is based on written or printed words requires reading, not listening; reading of text is interfered with by spoken words, even when the words are the same as those on the screen, except when the passages are very short. This is inevitable because people have different reading speeds and are usually ahead of the speaker, simultaneously trying to read while hearing different words.

Another point to be made in relation to this taxonomy is that there are very few communication systems that use only one means of communication, or one medium directed to a single sense. Books frequently contain illustrations; silent films depended heavily on printed subtitles and musical accompaniment; films and television contain printed words and still or moving images and, because of their multimedia nature, all of the possibilities of sound as well. Only the audio media (radio, telephone, tape, and so on) and the alphameric media (telex, twx, teletype, and other telemetry) are frequently used in their pure and simple form. These are, logically, the word media, based on (1) the spoken word and (2) the written word.

The first audio-based systems to be described in this section are examples of the audiovisual media combination we call television. Because television is so ubiquitous in western society, it is the medium that most people think of first when they hear the word "teleconferencing."

Post, telephone, and telegraph (PT&T) services in several countries have developed teleconference systems by means of which joint meetings may be held with participants located in two or more cities, even internationally. Teleconferencing is thus offered as another post office service along with telephone, telex, data transmission, and the like. This service is usually available in the two-way (binodal) mode, although some systems are able to incorporate three or four locations into a single meeting. The main marketing argument for these services is the lower cost of electronic meetings against face-to-face meetings with their attendant travel. The customers of these systems use them for occasional meetings or other special purposes, but only rarely for regular or routine meetings. It is the common experience of these systems that public acceptance is slow, and what customers there are, for the first year or so, tend to be first-timers. This is despite the fact that reactions are predominantly positive after first use.

Apparently inconvenience is a large factor in the low use of these systems; when there is but one teleconference facility in a whole city, it can seem inconvenient to go to another building, sometimes across town, even though the time involved is only a small fraction of that required for traveling to a distant location for a face-to-face meeting. With time, however, usage appears to increase. The current AT&T Picturephone Meeting Service, for example, seems to be fairly heavily patronized; after ten years of operation the British Confravision system was about 75 percent utilized.

In Chapters 2 and 3, the so-called audio systems will be discussed. We say "so-called" because most audio systems are as fully audiovisual as television systems, except that the visual portion is not in full and realistic motion. Nearly every audio system has some graphic element, for use when information can be made completely clear only if it is visualized in some way. In reality, such systems are *multi*media systems and should not be thought of as audio media only.

A few audio systems have a continuous visual element. Most, however, use visual materials only when they are needed. In some the visuals are "on line," which means they are transmitted by some complementary medium along with the audio component. In others the visuals are produced in advance, one set for each participating group, and distributed by mail. Slides, hard-copy materials, and overhead transparencies are the most common visual materials used in this way. In most cases the visual elements are one-way, while only the audio is interactive. This is also true of many interactive television systems — the television is one-way only; the audio portion alone is two-way and interactive.

In addition to this great variety of visual augmentation, some audio systems are also augmented by automated means of speaker identification and automatic switching. The various nodes (locations) of a teleconference system may be interconnected by a variety of different network configurations. Each of these variations has an effect on the way in which a system is used. Here again we see the fallacy of generalizing across all audio systems, or all television systems, or all alphameric systems. If one audio system is not effective for a given purpose, the conclusion cannot be drawn that therefore all audio has been tried, and found wanting, and that no audio system of any kind will be effective.

The alphameric systems, often called "computer conferencing" or "computer mail," will be described in Chapter 4. These are the least expensive, especially over long distances, and are also probably the least personal of the interactive media. They are sometimes interactive in real time, that is, people type back and forth to each other, words from each communicant appearing on all terminals. Mostly however, they are interactive in nonreal time, meaning that the separated communicants are not both communicating at the same moment. In this way it is like written correspondence, except that the delay between the sending of a message and the reception of its reply is more likely to be a matter of minutes or hours than one of days or weeks, as it is with letter writing. What we have called "real time" and "nonreal time" are often referred to as "synchronous" and "asynchronous" interaction. These terms will also be used in the following pages.

The fact that the delay between messages is so much shorter using alphameric systems than if they were written out and carried by post is very significant. Computer mail fills the large gap between telephone and postal correspondence, having some of the advantages of both yet none of the disadvantages of either.

The descriptions of example systems given in the following chapters are purposely brief and factual, without evaluation. The reader may wish to review them again when he or she encounters the discussion of the various elements of system design and its relationship to function.

1 TELEVISION SYSTEMS

Confravision

When it was installed in 1970, Confravision was the most elegant and expensive system so far built for video teleconferencing. It was conceived as a service of the British Post Office, available to the public at a fee. At first linking several cities in Britain, it was later extended to Sweden and Holland as well. The purpose was to provide a means whereby business and government executives could hold meetings via television rather than traveling. Two studios were built in London, and one each in Manchester, Glasgow, Birmingham, and Bristol. Studios in Stockholm and Malmo in Sweden and Amsterdam and The Hague in Holland were later added.

The standard Confravision studio was specially designed jointly by the technical staff of the post office and a leading industrial designer, each of whom made concessions to the requirements of the other. The resulting decor was most attractive, and this raised the question in some minds as to what effect such elegant industrial design might have on the users of the room and the quality of communication that resulted. It might be that such environmental factors are very important, or it might be, as someone had remarked in another connection, that there was seemingly more concern with the design of the bottle than with the quality of the milk.

Because the system was intended for the use of corporate executives, it was assumed that proprietary or confidential information might often be transmitted. To assure users of privacy in this regard, the system was designed to be operated entirely by the users themselves. Thus a sense of privacy similar to that provided by the telephone was obtained. A local studio technician was available in case of trouble and for maintenance, but he or she was visibly absent (sitting in the reception room chatting with the receptionist) while conferences were in progress. Standard television circuits were used for transmission between cities, but these were "backup" lines for those carrying the BBC and ITA broadcasts. They were consequently subject to occasional preemption, but this was not reported to be a problem; there were only one or two cases of preemption in the ten years of operation.

A moment's reflection on the part of any user would remind him or her that the post office must have dozens of technicians between London and Glasgow, for instance, whose duty it would be to monitor television transmissions, just as it must be in telephone service; but these are invisible and presumably neutral and trustworthy. It was necessary only to remove the visible technician to maintain a sense of confidence.

The standard Confravision studio contains two cameras, one for the participants and one for graphic materials. Two monitors are placed close to the first camera (so that when participants look at the monitors they appear to be looking into the camera). One monitor shows the distant group, and one the local group — the output of the local camera.

One set of push buttons at the chairperson's (center) location provides very simple camera and switching controls. The camera has two lenses on a turret, which can be changed mechanically between a wide shot, showing all five participants, and a medium shot, showing only the center three. Another button selects which of the two cameras will be used. The second camera is for graphics only, and is mounted on the ceiling above a small end desk that

Figure 2: Confravision system meeting room, British Post Office, London.

forms an L with the main desk. If and when graphics are used, an operator is supplied by the client. The operator changes the graphics, controls the picture area by zooming the lens, and points to elements under discussion as required. A small preview monitor to show the output of the graphics camera is provided at the graphics desk (see Figure 2, foreground). This very simple series of controls provides only limited visual possibilities, but it was considered necessary if totally inexperienced people were to operate the system.

Still one more desk position is provided, to the left of the graphics operator, for a secretary. This person may keep minutes, assist the graphics operator, and operate a built-in audio recorder if a record of the meeting is needed.

It is interesting to note that this system is an example of an *unswitched* system. The only switching that is done is local, between the conference camera and the graphics camera. There is a continuous feed from each participating location (in the case of a three-location conference) to each other location. This means there must be three television channels terminating at each location: one outgoing and two incoming. The same is true of the audio transmission that accompanies the video in a television system. Audio is thus "open." Anyone can speak at any time and be heard by all.

Without the elaborate existing distribution facilities of a post and telegraph service, a private three-location television system of this kind would have to install two incoming and two outgoing lines at each location; a four-location system would require six (see Figure 3). The cost of interconnecting multiple-location unswitched systems increases roughly as the square of the number of nodes, rapidly becoming astronomical.

While Confravision was originally designed for binodal conferences, within ten years the three-way conference had become the most common mode of operation. To maintain an unswitched system while operating in the three-way mode, each studio had to be equipped with a second incoming monitor. This was arranged by setting up a portable monitor on a wheeled stand somewhat over to the side. Had this side monitor been used to show the third location, eye contact with the people there would have been lost. At the third location participants would have seen people at the first location looking off to the side, rather than directly at them. Accordingly, this monitor was used instead for the local picture, and the central monitor associated with the camera (that was originally used to display the local picture) became the monitor for the third location.

In addition to the fixed studios, at least two portable units were in operation by 1980, one on loan to an industrial firm and the other used at fairs and expositions to promote the use of the service.

Among the users of Confravision could be counted banks, architectural firms, builders, engineering companies, makers and distributors of pharmaceuticals, insurance companies, unions, solicitors, and hospitals. On occasion, ailing animals in isolated zoos were diagnosed by veterinarians in London.

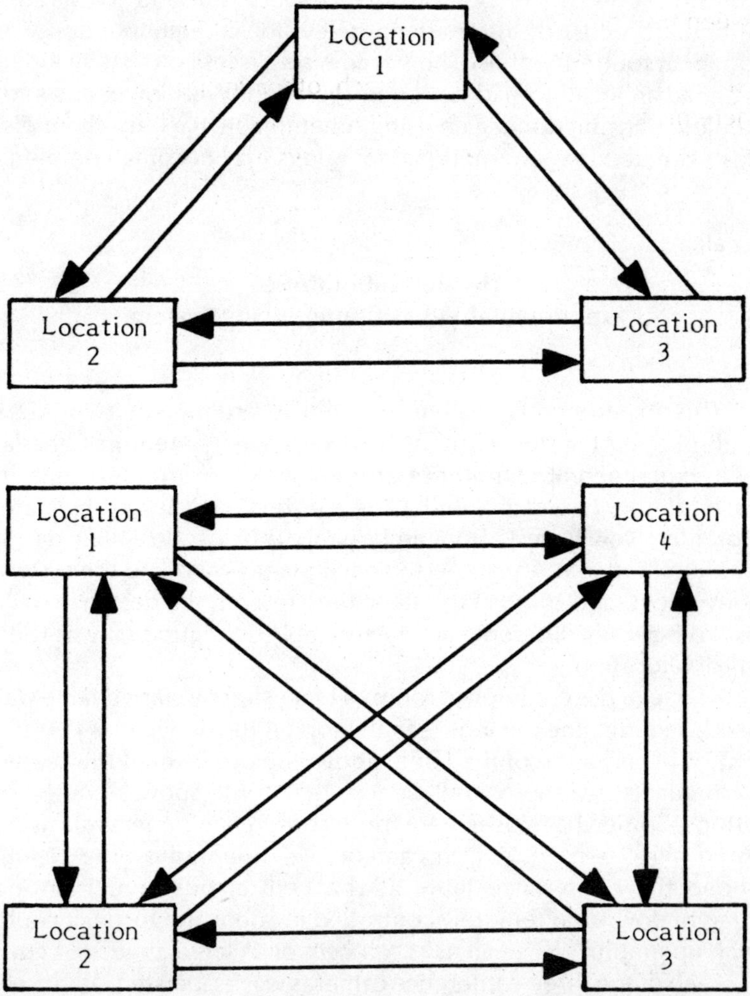

NOTE: A three-node network requires three links composed of six one-way channels, four channels terminating at each location. A small increase in nodes raises the transmission channel requirement greatly. A four-node system requires six links composed of twelve one-way channels with six channels terminating at each node. The formula is as follows: channels at each node = N (nodes) − 1 × 2.

Figure 3: Unswitched systems and the number of linking channels required.

In early 1980 the system was estimated to be about 75 percent utilized, sufficient to be considered successful and to be continued. The cost of international television interconnection was too great for frequent use, however. A single session that included a Canadian location had not been repeated. With lower costs for satellite transmission, again using preemptable back-up channels, it is expected that international meetings may become common.

The Bell Laboratories Experimental Video Conferencing System

This experimental system was the forerunner of two AT&T multiple-point systems and at least one such system in Canada. Two experimental conference studios were constructed, one in each of the two main Bell Labs Centers, at Murray Hill and Holmdel, New Jersey, about an hour apart by car on a busy day. It was assumed that researchers in each plant would find the system convenient, and could at the same time test out the design in daily use. A basically different camera and table configuration was tried in each location.

In one of the conference rooms a long, slightly curved table was used, seating nine persons (see the top picture in Figure 4). At the center of curvature of the table, above the eye level of the seated participants, was an overall camera that could show all nine positions. Somewhat closer were three conference cameras showing three persons each. A fifth camera was mounted on the ceiling above the conference table at the central position. Pointing straight down and remotely controlled in zoom and focus, it could pick up graphics as small as 11 × 15 cm or as large as 46 × 61 cm.

Each of the three conference cameras was associated, as closely as possible, with a large, high-quality, standard broadcast television monitor. The three closest monitors were approximately seven screen widths from the participants and presented the incoming picture from the other terminal at all times. The outgoing monitor, associated with the overview camera, was about twelve screen widths distant.

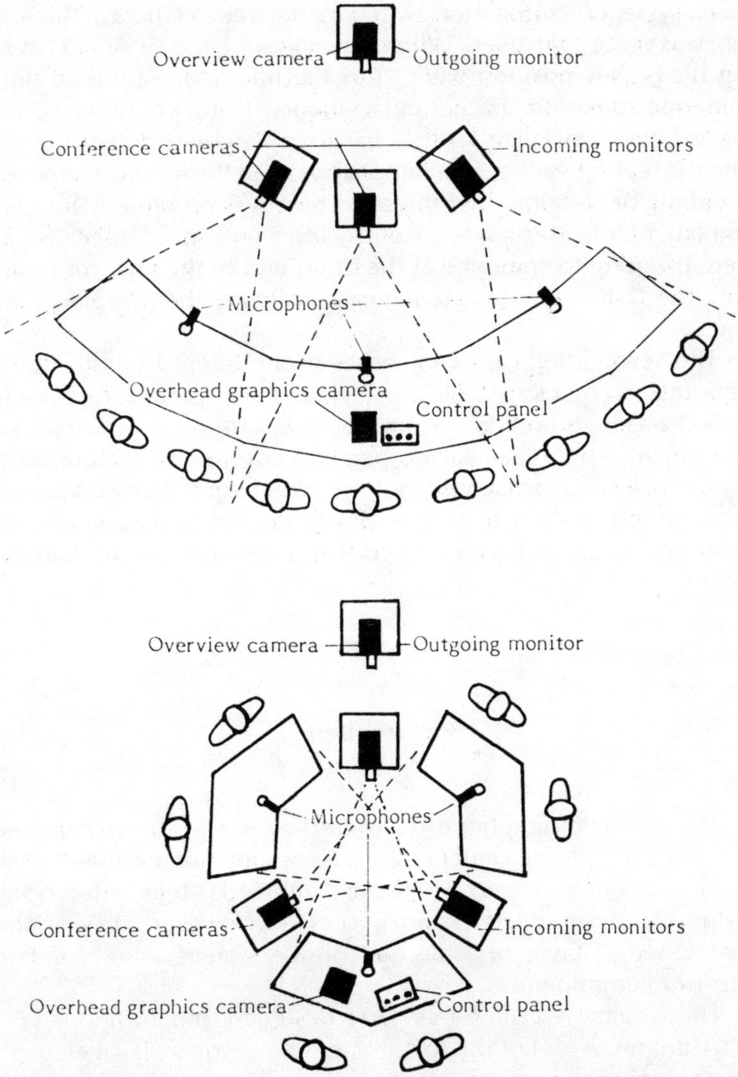

Figure 4: Bell Laboratories experimental video teleconference system configurations.

An innovation later continued in the systems designed by AT&T was a system of automated switching between cameras. This system was voice controlled. When a person spoke, the camera covering his or her position went onto the line and continued until someone spoke on another microphone. (Cough buttons, spring loaded, were installed so that participants could defeat the automatic feature so they would not appear on the line when merely coughing or clearing their throats. There was no remedy for general laughter.) After a few seconds, if no one spoke (because all were listening to someone at the other end of the line, for example), the system would automatically switch to the overview camera.

The second Bell Labs conference room differed slightly in design. Instead of a long table at which all participants faced essentially the same direction, a round table was used, across which six participants could face each other, two covered by each of three conference cameras (see the bottom picture in Figure 4). Viewing distance between participants and camera-monitor assemblies was about the same. In each case loudspeakers were mounted in the ceiling.

Picturephone

The name Picturephone, registered by AT&T, was first applied to a video telephone camera-monitor combination designed to sit on a desk top. With several of these combined, a teleconferencing system was designed. It was not successful, however, and was later abandoned in favor of a more elaborate system using standard television equipment.

The original Picturephone was designed and developed by AT&T on the model of the telephone, to provide a visual element for the enhancement of telephone conversation (see Figure 5). Video telephones have also been developed in Sweden, Japan, the United Kingdom, the Federal Republic of Germany, and other

Figure 5: Bell System Picturephone unit.

countries. These vary in picture quality, design features, and transmission bandwidth. Some are black and white, some are color, and some transmit a continuous series of still pictures, one every 30 seconds or so.

The terminal hardware of the Picturephone system was intended for the use of a single individual, two at the most, and thus it was made relatively small and compact. A 5.5-inch-wide monitor and a small camera were incorporated into a single unit, which was intended to sit on a desk-top pedestal within only two or three feet of the user. A microphone was originally incorporated into the same set, but for most applications it was found that the standard handset telephone, which was always on the desk anyway, served best for audio interchange. In the late sixties and early seventies it was expected that a video telephone service would be accepted, at least on an executive level or within large well-heeled organizations. However, the expected reception to AT&T's market trials did not materialize. Those who acquired Picturephone equipment found that they were severely limited in whom they could contact visually, and found little real use for the medium.

Picturephone[1] Meeting Service, First Generation

As an alternative approach, AT&T built several individual Picturephone sets into a teleconference system, and for a market trial installed teleconference rooms in four cities. (Two of such rooms were installed in New York.) As in the case of Confravision, these systems were intended for public use, at a fee. They were, like individual-office Picturephones, little used, but more for the reasons that Confravision was little used. With only one or two locations in a large city where a visual conference call could be made, people tended to feel that the inconvenience of scheduling a trip across town for a group of people, plus the premium cost, was too much to pay for the extra advantages of visual interaction; they preferred other less expensive and more convenient means, such as telephone conference calls or private audio systems. They also opted for full-fledged social interactive activities that no teleconference system, however sophisticated, can duplicate.

Although the first-generation Picturephone teleconference system described here is no longer offered to the public, it is a distinctive kind of system that has certain advantages and may be used in the future for private systems when both terminal equipment of various kinds and transmission channels are less expensive.

AT&T used a standard conference room and table with a maximum of six persons seated in the normal fashion on both sides of the table. Participants were seated in pairs — four on one side and two on the other (see Figure 6). Each pair faced an individual Picturephone unit that was just beyond arm's reach. Thus there were three "conference cameras." In addition, a fourth camera was used for graphics and a fifth provided an overall view of the entire table. The last two were simply cameras, not Picturephone units.

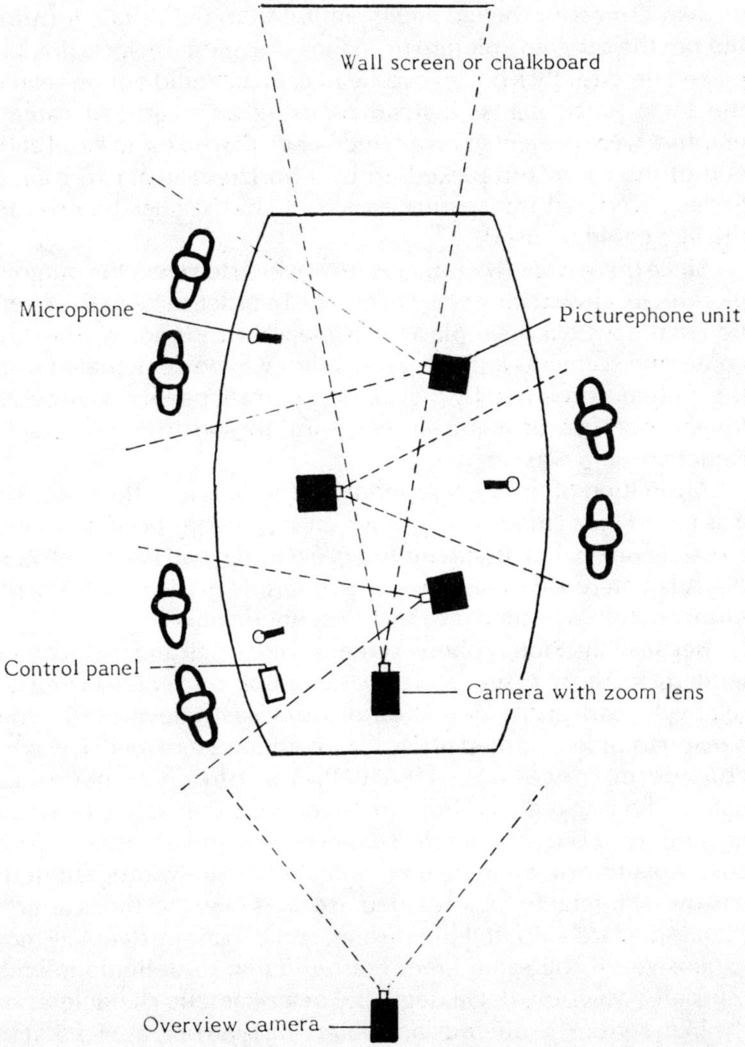

Figure 6: AT&T Picturephone Meeting Service, first generation, using table-top Picturephone units.

Since the Picturephone set can show only the picture taken by its own camera or the incoming picture from the distant location, but not the outgoing picture from the entire system, local graphics picked up directly from an overhead camera could not be seen by the local participants. Instead of using an overhead camera, graphics were projected on a screen or displayed on an easel at the end of the room and picked up by a horizontally aimed camera. Slides, overhead transparencies, easel charts, chalkboards, and the like could be used.

Since there were five cameras from which to select the outgoing picture, local switching was necessary. In order to keep the operation of the system as simple as possible, switching among the three conference cameras and the overall shot was voice-actuated using the system developed by Bell Labs. A control panel was provided for the purpose of manually switching to and from the graphic camera. Audio was open.

A variation of this Picturephone system was also tried out. This was based on a circular table at which six persons occupied a little less than one half of its circumference (Hough and Panko, 1977: 14, 16). All conference cameras were on one side of the room. The rest of the system was similar to that described above.

Because the Picturephone screens were small and their content limited, in most cases, to large close-ups of people, even the relatively low-quality definition of broadcast television (in comparison to slides, printed photos, and other picture media) was not considered necessary. Hence Picturephone transmission bandwidth was reduced from the standard 6-megacycle television channel to a one-megacycle channel, thus greatly reducing the cost of its transmission. Even this reduced bandwidth, about the narrowest tolerable for any video use, was over 200 times as great as the standard voice telephone line, and hence substantially more expensive. At the same time, the reduction in definition, which translates as visible detail, legibility of alphameric characters, and the like, seriously suffered and limited the type of graphics that the system could accept. The use of standard visual aids on screen or easel happily fitted this limitation, since most designers of conference materials have learned to make their designs and lettering very simple and posterlike.

Picturephone Meeting Service

This system was installed in 1976 to replace the Picturephone service previously described. It resembles Confravision more than any other system, although it is more sophisticated in several ways. Six or seven cameras are used in each conference room, and four groups in four different cities may be interconnected in one teleconference. In early 1976, when the market trial began, there were Picturephone Meeting Service conference rooms in four American cities; by 1979 the number had grown to eleven (Boston, New York, Philadelphia, and Washington on the East Coast; Pittsburgh and Detroit in the Northeast; Chicago in the Midwest; Atlanta in the South; and San Francisco, Los Angeles, and Sacramento on the West Coast). The Los Angeles, Sacramento, and New York City meeting rooms had color equipment.

This Meeting Service differed from the previous AT&T first-generation Picturephone conference system in that full channel (6MHz) television transmission was used, with standard closed-circuit equipment. All that remained of the original Picturephone system was the name.

From the experimental system at Bell Labs several features were incorporated. There were three conference cameras showing two persons each, and individual microphones that automatically controlled the video switching. When a person spoke, he or she was immediately shown in two-shot on the screen. When no microphone received sound above a certain level for a certain period of time, a general shot including all six participants appeared. In the Chicago facility this general shot was a multi-image frame composed of the three shots from the three conference cameras multiplexed into one frame (see Figure 7 for depictions of both types of overview shots). Bell Labs' individualized monitors (one for each group of three people) were not used; only one set of monitors was provided for all.

This system followed Confravision in many respects as well. Like Confravision, this was a public system, available to any users at a fee. Also, it was operated entirely by its users; there were no room attendants or technicians present during conferences.

a. Overview shot as produced in Los Angeles Picturephone meeting room.

b. Overview shot as produced in Chicago meeting room. (A composite of the three main conference cameras.)

Figure 7: Picturephone Meeting Service (second generation), overview shots.

Furthermore, the Picturephone Meeting Service rooms were very attractively designed. An oval table rather than a straight table was used, with the six participants seated along one side (see Figure 8). As in most video systems, the graphics camera was mounted above, so that it looked down on the materials. As in Confravision, it was separated from the conference table. The graphics display was operated either by a special person or by one of the participants, who had to leave the table to operate it. The graphics stand contained two small monitors (Figure 9). One showed the operator what he or she was doing with the graphics, and the other showed the incoming picture. Thus the operator could see and talk to a questioner directly. A backlight was provided beneath the table top (Figure 10) so that transparencies could be used.

The system provided more than a graphics camera for auxiliary visual materials. In most of the facilities, there was a slide chain (projector plus camera) installed in the cabinet below the graphics camera. The projector was an Eastman Carousel, the most commonly used type. The user simply brought in a tray loaded with slides in the correct order, and he or she could advance them from the control panel in the center of the conference table (see Figure 11).

In addition to this, all meeting rooms had a camera mounted on a tripod that could be moved freely about the room to pick up any other kind of picture that might be desired. This could be an overall room shot, for instance, or a close-up of a chart or drawing board on which easel charts could be displayed or on which drawings made on large sheets of paper with marking crayons, or brown or green chalk boards, could be used. This extra camera was available in the Los Angeles meeting room also, even though an easel camera was built into the wall at that facility (see Figure 8, right-hand photo).

Two main monitors were used, one showing the local output and the other the incoming picture from a distant location. In some of the Picturephone Meeting Service rooms these two monitors were mounted side by side, in the manner of the Confravision studio. In some they were one above the other, local output at the top. In the case of a three- or four-way switched conference, a means was necessary to identify the source of the picture appearing on the lower monitor. This was done with a series of three translucent panels below the monitor, into which nameplates of the two or three distant locations could be placed to light up automatically and identify the source of the displayed picture.

Finally, a videotape recorder was provided so that the entire conference could be recorded for the benefit of persons unable to attend, or for a permanent record (Figure 12). For use during the conference a hard-copy printout machine (Tektronix Model 4632) was provided. The subject at the other end must be held still for five seconds to make an exposure by this machine; in five seconds more an 8½″ by 11″ print is ready. In short order all participants could be provided with their own copies of what they just saw on the screen.

a. The graphic stand is seen in the background, left of center. A drawing board on a wall-mounted easel is also provided, far left, for the use of standard large-size conference materials.

Figure 8: Picturephone Meeting Service facility in Los Angeles.

b. The camera that covers the drawing board was built into the opposite wall and equipped with a 10-to-1 zoom lens controllable from the center position at the conference table, or at the easel. No pan or tilt controls were provided at the central control console for this easel camera. Only a zoom control was provided; zooming in produced a closer shot of the center of the field only. However, next to the easel itself, for the use of someone standing at the board, was a joy stick that could be used to show a close-up of any part of the board.

Figure 8: (Continued)

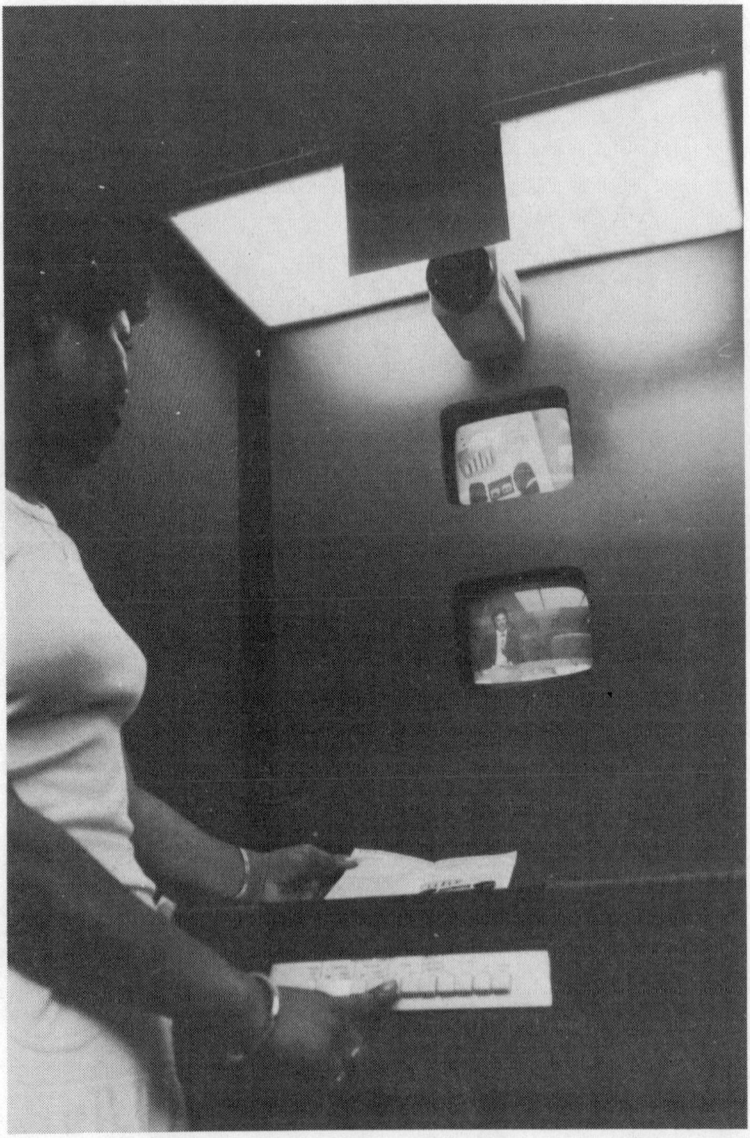

NOTE: The upper monitor displays a preview of the graphics camera. The lower monitor shows the incoming picture, on which the graphics operator may see the distant group and talk with them even though they cannot see him or her.

Figure 9: Typical Picturephone Meeting Service graphics stand and camera.

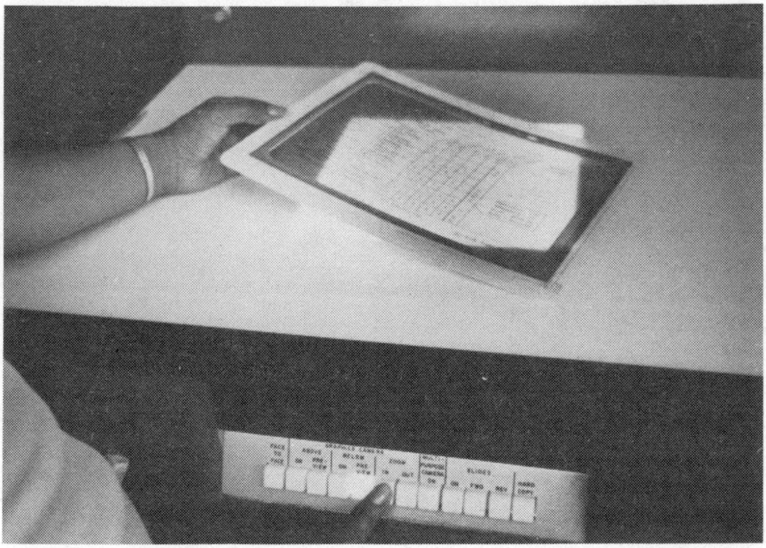

NOTE: Controls include above and below lighting and zoom in and out (1-10 zoom), plus a duplication of several key controls from the central table control panel so that the chairperson can leave the table and operate graphics while still controlling the meeting.

Figure 10: Graphics stand table, with rear illumination for transparencies.

In addition to these auxiliary devices, the Los Angeles facility and some others added another service — facsimile. Users could exchange typewritten materials and other printed or graphic materials either before or during a meeting. A four-minute facsimile machine was provided, along with the rest of the service, and there was no line charge for the distribution of materials. The facsimile machine was installed in an adjoining room.

While Confravision, both in Britain and in Australia, was very slow to prove popular, the Picturephone Meeting Service in America that started up a decade later was busy daily within a year of its inauguration. In late 1979, the Los Angeles facility served some six to ten customers a week, with the average meeting running two hours. Four-hour meetings were not uncommon. At the rate of $6.50 per minute (in 1979) between Los Angeles and New

NOTE: Controls are as follows: VOLUME—controls loudspeaker level (it is not possible to raise the volume louder than the feedback point); AID—rings a buzzer in the anteroom to summon a staff person; FACE TO FACE—activates the three conference cameras (while this button is lighted, switching is automatic among these three cameras, determined by which microphone picks up the loudest sound); OVERVIEW OFF—overview camera is *not* switched automatically if no sound is picked up for 20 seconds; OVERVIEW AUTO—operates overview automatically; OVERVIEW ONLY—the wide shot is switched on and stays on until further notice; MULTIPURPOSE CAMERA (either the camera built into the wall that covers the graphics easel or the extra tripod-mounted camera); PREVIEW—the camera is turned on and its picture is shown on the monitor; MULTIPURPOSE CAMERA ON—this camera feeds the outgoing line; ZOOM IN—lens slowly enlarges the center of the field; ZOOM OUT—field slowly gets wider. (Both zoom buttons are spring loaded; the action stops when the button is released. In Chicago, where the tripod camera was manually operated, the last four buttons were absent.); GRAPHICS ABOVE—graphics camera and top lights are turned on; GRAPHICS BELOW—graphics camera and rear lighting (below table) are turned on for transparent materials. (The graphics camera can be put on the line either from the conference table or from the graphics stand. However, the only preview monitor for this camera is at the graphics stand.); SLIDES ON—slide projector lighted and slide pickup camera put on the line; SLIDES FWD—advances slide projector; SLIDES REV—reverses projector to previous slide; HARD COPY—activates the Tektronix hard-copy machine.

Figure 11: Control panel inset at center conference position.

York, a two-hour conference would cost the customer $780, about the same as the round-trip travel cost for only two people. In addition, attending a meeting on the opposite coast loses the traveler a day's work, because of the west-to-east travel, and two days of per diem cost are incurred. It is mainly on this basis of cost

NOTE: The VTR can be seen in the foreground; the Tektronicx hard-copy printout machine appears in the background.

Figure 12: Recording equipment at side of meeting room.

comparison that the use of this system has been justified. Whether a company, in the long run, actually cuts back on travel costs after beginning to use teleconferencing is another question, however. Most Picturephone clients were first-time users; regular users were slow to develop.

Other Video Systems

A few more video systems should be mentioned, at least briefly, since they incorporate some features not found in those described above. The Australian and the Japanese PT&Ts have used a system that increases the information that can be carried on a single video channel. A brief description of the Australian system will explain both.

Australian Confravision

Like the British Post Office and AT&T, the APO (Australian Post Office) developed a public system linking several cities, but only two at a time. Privacy was a concern of this public system also, and it was designed to be operated by its users without monitoring by an operator or technician.

The system could accommodate six participants at a curved table, using two conference cameras and one camera for graphics. Each of the conference cameras could show three participants. Wanting to avoid either manual or automatic switching between the conference cameras, the designers of this system made it possible to transmit both cameras at all times. Using a second television channel was not a practical way to do this, since the cost of transmission over even one channel is the greatest single cost item in any long-distance video conference system. The solution was to multiplex the two cameras into one transmission channel.

A shot of three people seated along a table (showing each person from head to waist) normally contains large areas of fore-

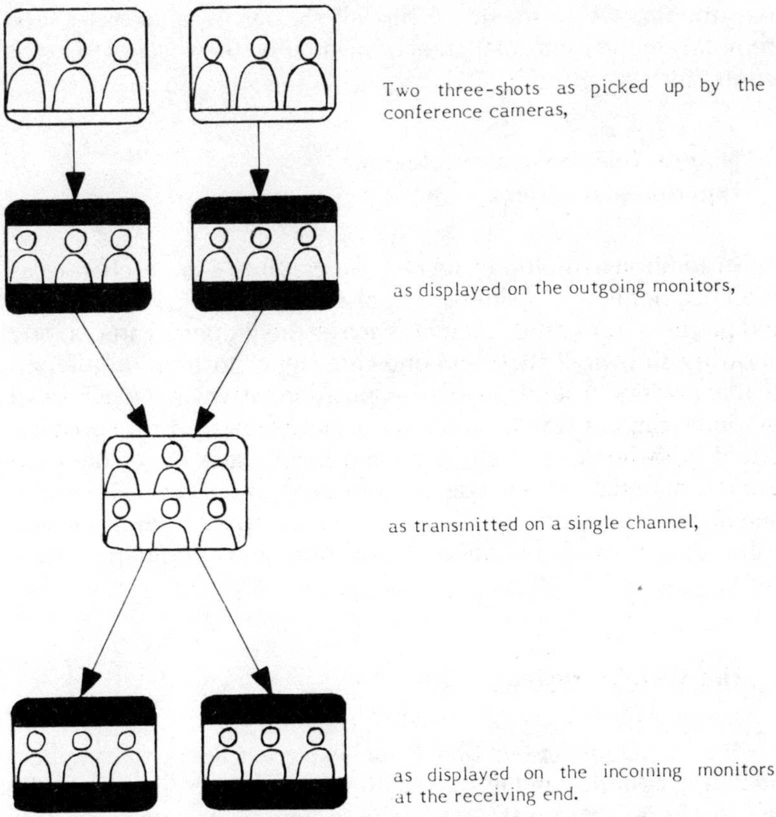

Figure 13: Australian Confravision multiplexing system.

ground table and/or background above the people's heads that is essentially waste space. The APO system reduced the height of the used picture area on each of these conference camera shots to one-half the normal amount, cutting out only the waste space (see Figure 13). Instead of the standard aspect ratio of 3 units high to 4 wide, these APO cameras produced, in effect, pictures that were 1.5 units high to 4 wide. Multiplexed, one above the other, they constituted a full raster and were transmitted that way on a single channel. On reception they were again separated and displayed on two monitors side by side. Thus all six participants were seen remotely in the same spatial relationship that they occupied in the conference room.

Nippon Telephone and Telegraph Experimental System

In addition to multiplexing two cameras into a single channel as APO had done, the Japanese system also simultaneously transmitted pictures from other cameras such as the graphics camera, one showing an overall shot, and one showing a close-up on the particular position (called the lecturer position) at which the overhead graphics camera was located. A speaker seated in this position could both be seen in close-up and manipulate his or her own graphic materials. NTT was apparently experimenting in two opposite directions at once, the saving of channel space by multiplexing cameras and the simultaneous transmission of additional channels.

The MRC-TV System

The Metropolitan Regional Council, a council of governments, including counties and municipalities in the New York metropolitan area, was dedicated to improving communication and cooperation between governing bodies (of which there are some 1200 at various levels within a 50-mile radius of New York City). Because travel in this area was particularly time consuming, it was thought that an interactive television system would make intergovernmental meetings possible where they had not been so before, while extending their range and usefulness.

Starting with ten locations, from each of which audiovideo response was possible, the system grew from nine to fourteen locations, extending from New Brunswick, New Jersey, on the south, to Rockland County on the north, and as far east as Stamford, Connecticut.[2] This was a switched system, and the only full two-way video system in which participants at more than two or three locations could interact with each other in the same meeting. In most of the MRC-TV meetings, most of the locations participated; meetings in which all locations participated were by no means rare. If 10 people took part at each location, a meeting could involve 140; there have been meetings in which over 200 participated.

While the system was designed for meetings, it was found that courses in clerical skills, supervision, management, and many other subjects were badly needed in outlying counties, for which the system proved to be a perfect vehicle. This led to the development of a kind of interactive teaching, which was very different from one-way instructional television, resembling Socratic teaching in the classroom much more closely.

MRC-TV began operation in 1973 and continued until 1979, when the Metropolitan Regional Council itself closed its doors. The system had been in daily use from 10 a.m. to 4 p.m. for scheduled meetings and classes, and beyond those hours for on-demand playback of videotape recordings of previous meetings. About 35 percent of the system's use was typically for meetings and 65 percent for interactive instruction. This was on-the-job training for government workers who, but for the TV system, might not have had such opportunities for self-improvement at all.

The MRC-TV system was designed by a firm called Genesys, which had previously designed and built instructional television systems for Stanford University, the University of Minnesota, and others. These earlier systems had all been characterized by one-way video and two-way audio, making it possible for advanced engineering students located in industrial plants throughout a metropolitan area to receive video lectures and to interact with the instructor and with students at other locations via audio.

Transmission in the MRC system was by microwave, not by wire, as in the case of the previously described systems. A single relatively powerful omnidirectional transmitter fed all locations from a central point atop the World Trade Center building in downtown New York. Each location was provided with its own less powerful transmitter and a parabolic antenna to feed the return signal into a narrow beam aimed directly back to the central point.

Because there was only one main channel, carrying the same picture to everyone, this system was, like most such systems, limited to only one conference at a time.

A control room and a studio were located at the central point. The control room contained switching, audio, and special effects equipment for the entire system, and also for the associated studio. The central studio was thus one of the fourteen locations. The output of any could be switched onto the main outgoing channel and reach all participating locations. All participants could see and hear a speaker at whatever location he or she might be. Because the system had only one outgoing channel, however, a speaker anywhere but the central studio could see only him- or herself.

This condition was not conducive to good interaction. A speaker needs to see at least some person or persons to whom he or she is speaking. Fortunately, an inexpensive special effects system had been incorporated with which a limited number of split screen effects could be achieved. Such a device allows two locations to be displayed on the same screen, in various areas and proportions. The screen can be split by a vertical line or a horizontal line, or the two lines can be combined to put the picture from one location (a close-up of a speaker, for instance) in a corner of the screen, while the picture from another source (a questioner or a listening group) occupies the remainder of the frame. Since both these lines are movable, the area occupied by one picture may be gradually extended while the other is reduced. Thus one area may be "wiped" into the other or out again. This apparatus was an inexpensive version of a standard part of nearly all television studio switching gear, which has little real need in one-way television. In an interactive system that has only one distribution channel, however, it is indispensable — hence this extended description. One of the characteristics of full interaction, as we saw in the Introduction, is the provision of constant, simultaneous feedback. A speaker must see, or hear from, the person he or she is talking to. Without this, at least for the time he or she is talking, the speaker may as well be using a one-way medium.

To provide simultaneous feedback, the MRC-TV controller often put a head-and-shoulders close-up of a speaker in a corner of the screen, surrounded by a shot of either the immediate questioner to whom the speaker was directly replying or an arbitrary selection of some listening group.

In contrast to the public systems (Confravision and the Picturephone Meeting Service), which simplified the choice of camera shots to a very few, this system provided a range nearly as great as a standard television studio. It was of course far more complex than a participant could handle by him- or herself while also conducting a meeting (a chairperson of a Confravision group of five, for instance). So in each of the ten conference rooms, a control console was provided. This, like the central switching equipment, was put together from off-the-shelf closed-circuit parts. It required an operator, and each participating government chose some lower-echelon person, often a secretary based in a nearby office, to be in charge of turning on and turning off the system, focusing the camera, and when needed, operating the camera throughout a meeting.

Camera operation was remotely controlled from the console. The operator could pan and tilt (left-right and up-down), focus for different camera-to-subject distances, and zoom the lens through a range of ten to one. This meant that the camera could show an entire room full of people, then, at a fixed rate, zoom in to one-tenth the width of view and single out but one participant to show in full close-up. Panning, tilting, and focusing were also required during this zooming maneuver. When done with skill this is routine and unnoticeable. When done without skill, as was usually the case at first, it was even occasionally irritating (the operator might pan left before discovering that he or she needed to pan right, make the focus worse before finding he or she was adjusting it in the wrong direction, lose the speaker completely for a moment or two, and so on), but the result was almost always worth the cost, because the speaker could be seen so much better in the eventual shot.

The same simple controls provided in each of the outlying conference rooms were installed in the main control center, which faced through a large window on the central studio. Thus the single operator at this center controlled two studio cameras as well as the central switching; but, because this operator was thoroughly trained and had daily practice, this rapidly became no problem.

Here again, occasional mistakes and inadequacies that would be considered unforgivable in broadcast television make very little difference in interactive TV. This is probably because so much one-way television is make-believe, and a sudden consciousness of technique can break the illusion. The production standards necessary for story-illusion TV tended to carry over into actuality TV where they were inappropriate. This was also partly a holdover from film production conventions, in which, for many years prior to television, showing a camera or a microphone in the shot was considered disgraceful.

Another feature was necessary in the design of this system that is not required in unswitched systems; each outlying conference room leader had a "request-to-talk" switch on the desk before him or her. This lighted a label on the central control board, and the central controller could inform the meeting chairperson (by holding up a sign) which location wished to speak; when the questioner was recognized, he or she was then put on the line, usually in split screen.

The Reading Pennsylvania Community TV System

The Reading system was built in 1975 as a National Science Foundation-funded experiment in the use of two-way cable television for community purposes. After a year and a half, when NSF funds came to an end, the system was continued with local funds. The senior author had some firsthand experience with this system as a consultant early in the implementation stage.

The primary part of this system consisted of three locations, which were called Neighborhood Communication Centers (NCCs). The ATC-Berks Cable TV Company had some unused channels, one in the standard downstream direction and, quite unusually, three in the reverse, upstream direction. All four converged at the cable TV "head end," the central control point for the parent cable system. The same location, with an additional operator, became the control point for the interactive system as well. Each NCC fed its own television signal to the head end, thus using the three upstream channels. At this location, switching equipment was installed to choose from the three feeds which picture and sound would go out over the single downstream channel to all three NCCs. In this way the Reading system was similar to MRC-TV; there was a single outgoing line that carried the same picture and sound to all locations. The only difference was that the MRC-TV transmission was by microwave, while in Reading it was by cable.

Each NCC was a gathering place for senior citizens. After the TV system was installed, a group of six or a dozen elderly persons at each center participated in entertainment and informational activities in interaction with others across town.

In addition to the three neighborhood centers, the same downstream channel distributed the interactive activities as a one-way program to all subscribing homes on the entire cable TV system. Viewers in these homes could become participants, if they wished, by calling in on the telephone. Reading is a city of about 88,000 population, with a higher-than-average proportion of senior citizens, some 14,000 in all. In the various studies of the system made by the Alternate Media Center of New York University, it was found that 1,000 of them participated at least once during the experimental period, either at the NCCs or via telephone.

Probably the most significant thing about this system was that it was operated by senior citizens, who, rather than specialists or personnel from any agency, determined the content of the sessions held. Socially, it brought them together for joint "sing along" sessions and quiz and game programs. Intellectually, it involved them in discussions of current social issues, political questions, and the like. Informationally, it gave them a regular opportunity to confront public officials of city and county government to request information, give their opinions and recommend action, request services, and learn more about the operation of their government. The mayor and members of the city council were glad to devote their time to this and became so well known to the television participants that they were often addressed by their first names. Red Burns, executive director of the Alternate Media Center, made this comment: "The mayor and council members originally thought of it as a way to talk to people, but they found out it was a way to hear what people have to say."

To make participation on the part of government officials as convenient as possible, a set of remote TV equipment was taken to their offices. This was regularly set up on Mondays at the county building, Wednesdays at the social security headquarters, Thursdays at city hall, and Fridays at various local high schools. Installations had been made at each of these locations to feed audio and video into the cable, as well as to receive cable signals.

A few local young people were hired and trained as part-time "technical aides" for the project, and the best of these were placed in coordinating positions when local control took over. The role of switcher at the head end was originally a full-time staff job, which was turned over to a previous technical aide. There was usually one such technical aide at each NCC during interactive sessions.

The technical aide in each studio operated one inexpensive black-and-white conference camera and could switch to another that was set up for graphic materials. The remote unit, which was carried to city hall and other locations, consisted of a single camera, with associated video and sound equipment. Because only one channel fed all locations (channel 3 on the cable system) the Reading System, like MRC-TV, was dependent on the split screen. Before the need for this was fully realized, one participant expressed the need very succinctly: "I want to see the person I'm talking to." Without a split screen he saw only what everyone else on the system saw, namely, himself. So that a participant could see the person he or she was talking to, the switcher split the screen vertically, with one participant in the left half and another in the right, and the lenses of both cameras were adjusted so the images were of equal size. When one of these was a little old lady and the other the mayor of the city, and they were equal in size on the screen and in volume on the loudspeaker, the democratic effect might have contributed to her tendency to call the mayor by his first name.

As in MRC-TV, the conference cameras were equipped with 10-1 zoom lenses and were manually operated. This provided a range of shots from a wide view including all participants to a head-and-shoulders shot of a single participant. The camera, of course, had to remain in one place, as close to the monitor as possible, so that in watching the monitor the participants would appear to be looking into the camera. Unlike in the MRC-TV system, the cameras were not remotely controlled by an operator at a console, but by a standing camera operator (the technical aide). The camera was placed at the side of the monitor and was probably at least a meter from the center of the screen.[3] In the reports on this system there was no mention of the presence of the camera operator being distracting; it was evidently accepted as part of the system. Nor was there any comment about the wide separation between camera and monitor, which must have given the impression that people were looking off to the side instead of directly at those to whom they were talking. This too may have been accepted and quickly forgotten.

In each NCC there were several microphones, and these were left open at all times — the "open audio" condition. To prevent acoustical feedback,[4] they were not set at a very high level, however, and had to be held within five inches of the mouth for the user to be heard. In order to speak, a participant had to first obtain a microphone, usually only a few seats away and passed from hand to hand. At the beginning many people forgot to do this and started speaking without a microphone. When a participant had a microphone and held it up but away from his or her face a foot or two, it was a signal to the camera operator that he or she wanted to speak soon, either after the present speaker was through, or when the present topic was finished because he or she wanted to bring something else up. When participants held the microphone up close to their mouths, it was a signal that they wanted to say something right away. When they saw the camera pointed at them, they knew they had a good chance of getting the floor, but that they could not yet speak because they were not yet on the line. When, however, participants saw themselves on the monitor, usually in split screen, they knew they had made it. They could then either wait politely, as most did, until the present speaker was finished, or, if they felt their comments couldn't wait, interrupt then and there.

In the early planning stages of the Reading system, as also in the planning for MRC-TV, there was considerable concern about maintaining control of a meeting. The interrupt feature, particularly, was a cause of this concern. It was felt that if anyone could interrupt whenever he or she wished there would be no limit on the number of people who might speak at once. Meetings could quickly disintegrate into chaos. In both cases, senior citizens in one and government workers in the other, the usual protocols of courtesy and politeness in group behavior carried over. People took turns and waited for each other to finish, or, when they did interrupt, did so in a helpful spirit. Of course, we have all experienced face-to-face meetings that have become so emotionally charged that tight control must be applied by the chairperson just to maintain order. Such events rarely happen in teleconferencing, perhaps because the range of conference tasks is not so broad that such highly controversial subjects and volatile situations are often encountered. Most of teleconferencing seems to be devoted to fairly dispassionate information exchange and problem solving.

The Back Channel

In another way also Reading was similar to MRC-TV; in each case there was a need for an additional two-way audio system for operational use. At Reading this was called the "back channel."[5] It did not carry any of the content of the teleconference. In American broadcasting it would be called a "private line" (PL). A PL is usually for the use of technical personnel for the maintenance and operation of the main system. Before the private line was established at Reading, whenever electronic adjustments were needed at a center, the cable TV engineer at the head end who was monitoring all output would simply have to come in on the regular audio line, overriding whoever was talking at the time, and interrupt the session.

A frequent need for a back channel is in setting up a split screen. Before the switcher at the head end can combine two pictures in split screen (with the simple special effects device described earlier), the questioner from one center and the speaker from the other have to be placed on opposite sides of the screen. The special effects device then combines the left side of one picture with the right side of the other (or top and bottom, or some other pair of complementary areas). Camera operators usually draw a line down the middle of the view finder in cases such as this, to help compose the shot in the half frame. However, the switcher must tell each camera operator which half to use, and direct him or her to zoom in or out, tilt up or down until the two people are of the same size and their heads are at an equal level on the screen. Both camera operators and switcher wore headsets with one earphone and a small microphone attached.

The MRC-TV system had much the same experience before installing a PL (back channel). Without the PL, contact between the central control point and a distant location was achieved via the regular outside phone. This did not in all cases reach the camera operator at the distant point; sometimes it was answered by a secretary a few doors down the hall and instructions had to be relayed. Emergency situations, such as acoustical feedback that was making the system unusable, could not be remedied quickly.

The Misar II System

With the increased use of video conferencing in industry came a demand for consulting and development services in this field. By late 1982 there were several competing firms offering design, development, and implementation of video teleconferencing systems. Many of these firms went little farther than the first attempts at adapting standard video production equipment to conferencing purposes. Misar Industries, however, developed a far more sophisticated system, including elements of hardware designed specially for the purpose.

The Misar II system (already a second generation), is compared in Table 4 (Chapter 10) to the other systems previously described. The comparison is highly favorable. To begin with, Misar II uses a projected image, rather than a direct-viewing tube, so the image is roughly twice as wide as in other systems. This means that more people can be accommodated in the meeting room. (Misar's floor plan shows eight people at the conference table and another five secondary participants seated behind them.) The large screen also makes possible shots such as the one shown in Figure 29 (see Chapter 10). in which the image is life size. Viewing distance to such an image is the same as its virtual distance (which means simply that the person seems as close as the screen; proximic relations such as this will be discussed in detail in Part III).

The core of Misar's design is in the manner in which the conference cameras are mounted and operated. Only two cameras are used. Instead of being positioned horizontally (either in full sight or behind a partition looking through a slot), Misar's two color cameras point upward and look out through their slot via small mirrors above the lenses. A camera is then "panned" and "tilted" (left-right and up-down movements) by movements of the mirror rather than of the camera. The elevation drawing (Figure 14) shows the camera (item 11), the optical block containing zoom lens and remote-controlled servo motors (item 10), with the mirror shown at a 45-degree angle, allowing the camera to look out just over the top

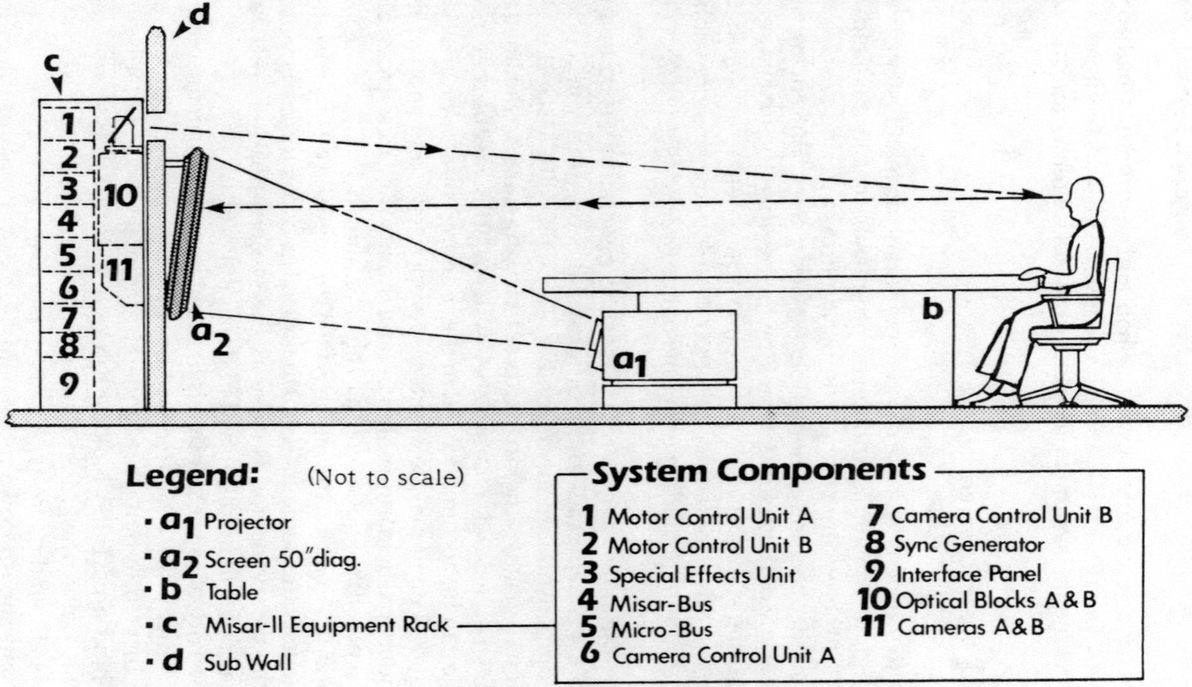

Figure 14: Misar II video conference system: elevation.

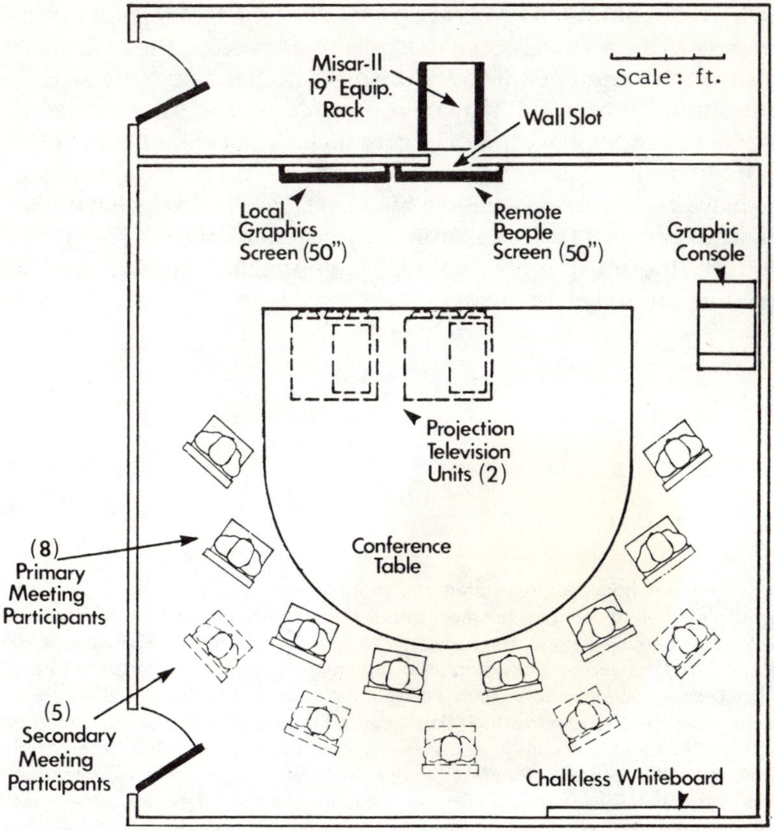

Figure 15: Misar II video conference system: plan.

of the screen. This allows a very small displacement between camera and viewing screen. From the average 14-foot viewing distance, the parallactic angle is only 5 degrees, about half that of earlier video conferencing systems (this subject is discussed more fully in Part III). The optical blocks are computer controlled and can be preset to cover any set of 175 possible room positions. The servo motors and the small light mirror assemblies allow simultaneous change of pan, tilt, zoom, focus, and iris controls within a fraction of a second.

In the most common mode of operation, switching is automatically voice controlled. When two people at the table speak simultaneously, the system produces a split-screen composite of the two shots so that both people are shown at once, and both life-size. A ten-button "Image Pad" may be used by the participants to manually select additional camera targets such as blackboard and other graphic devices, slides, videotape, and so on. Local graphics are displayed on a second screen (shown in Figure 15, to the left of the main screen). So far as we know, this system is, at the time of this writing, the most sophisticated development of automated TV production in the video field.

notes

1. "Picturephone" is a registered trademark of AT&T. It has been used by AT&T for three different services, however, and these are distinguished in this book as follows: When used alone, the name "Picturephone" will refer to the video telephone characterized by a desk-top unit containing a camera and a small monitor; "Picturephone Meeting Service, first generation" will refer to the now-abandoned teleconferencing system in which participants sat on both sides of a conference table and Picturephone units were used rather than standard television cameras and monitors; "Picturephone Meeting Service" will refer to the second-generation AT&T system in which standard television equipment was used (second generation will be implied).

2. As a staff member of the Rand Corporation, the senior author was assigned to this project for two years during its implementation and early operation stages. As a consultant, he helped solve problems, recommended improvements in system design and use, and analyzed reasons for success or failure (see Bretz and Dougherty, 1974).

3. This caused a parallactic difference in angle of view of 17 degrees at the average viewing distance, twice that of the Picturephone system (see Table 4, in Chapter 10).

4. See Chapter 2 for a discussion of this problem.

5. "Back," in this context, is meant in the sense of "back door" or "back stairs," not "backup." A backup channel would be a reserve channel ready for use in case the main channel should fail.

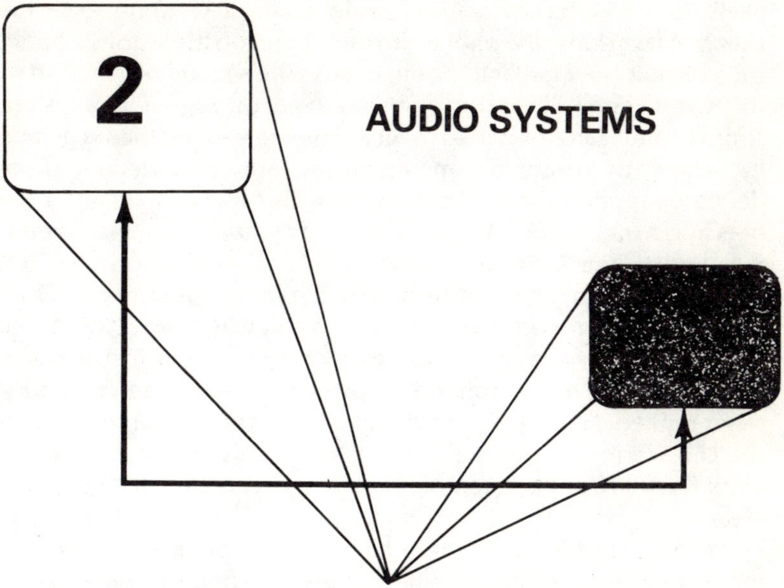

Plain Old Telephone Service (POTS)

Of all the interactive telemedia in use today, the telephone is by far the most familiar and the most ubiquitous. Even in relatively undeveloped countries, such as Burma and Bangladesh, where there is no more than 1 telephone to each 1000 inhabitants, there are probably few adults who have never seen one, although experiencing its use may be a different matter. At the other end of the scale, in the United States, Sweden, Canada, and New Zealand, where there is 1 phone for every 1 or 2 people, as soon as babies learn to talk they are urged to converse with Grandma on the phone. In West Germany, France, the Netherlands, and Norway, the phone is almost as omnipresent. The average in these countries as of 1975 was about 1 phone to every 3 or 4 people (Dolmatch, 1979).

In the developing countries, particularly those developing most rapidly, telephone service is plagued with technical difficulties. Often one must dial a number many times before the entire series of pulses goes through, and even when a connection is established it may not last for long. In offices in some countries, each secretary has two phones. To make a call one dials the number, lays down the phone and dials again on the second phone, then checks the first, dials again, checks the second phone, and so forth, until one or the other manages to get through. On occasion, simply going across town to meet someone is easier than reaching him or her by phone in some countries. Added to this is that the demand for phone service greatly exceeds capacity in many countries and a wait of several years after application is not uncommon. Substantial "fees" can reduce these waiting periods of course, but in Rio de Janeiro, as recently as 1978, the first author was told that the *official* fee for a telephone installation was over $2000. Once when one of the authors had expounded on the instructional applications of the telephone before an audience in Brazil, a listener replied, "I can see how the telephone could be very useful in the U.S., Señor, but in our country the telephone system doesn't work." While the statement was an exaggeration, the meaning was clear.

Where phones work, there have been some imaginative and innovative uses. In Israel, where there is a phone to every four people, the army had an interesting idea. On the theory that contact with home was important for morale and fighting spirit, phones were available for personal use, right up to the front lines. This interactive communication was considered worth whatever security risks might be involved.

Conference Calls

Wherever it was used, the telephone has been almost exclusively an individualized medium usable for one person to talk with a single other person somewhere at a distance. For many years, however, it has been possible, with the assistance of an operator, to tie together a number of people at different locations into one phone conversation (50 one-way locations or 38 two-way interactive sites, using 1979 equipment). Such a conference call must be scheduled in advance and hence takes precedence over unscheduled activities for each participant. When a large number of people are involved, it may take several calls to each before a time is found that is convenient to all. The shorter the notice, of course, the greater these difficulties, just as with any other kind of meeting. Most of us have experienced the difficulty of spontaneously organizing four or five persons at an office to go together to lunch all at the same time. Even after a satisfactory time is found to schedule a conference call, the problems are not over. At the start of a recent conference call in which one of us participated a dozen scheduled participants were gradually contacted by the operator and asked to wait. From time to time she returned to the line to report progress. The final and greatest problem was to locate the conference leader himself, whose phone line was apparently busy.

The development of electronic switching at telephone exchanges made a more informal, spontaneous version of conference calling possible. AT&T calls this the "add-on conference feature." A telephone user who has a phone with two or more lines feeding it (a row of buttons on the base) can usually be provided with this service for a few dollars extra per month. No operator is needed. If, in the middle of a phone conversation, you decide to include another person, you can put your present party on hold, switch to another line and dial the new person's number. When you contact him or her, you may use a special button to bring in the first person again on a three-way conversation. The first person may do the same if he or she wishes to bring a fourth person into the conversation, but this is as far, at last report, that this "chaining" can go.

In the case of an operator-arranged conference call, a "telephone bridge" is required. This device can make it possible for thirty people to be in open-audio interconnection. It has not been in frequent use, however. Amitai Etzioni, of the Center for Policy Research, reported setting up a teleconference in New Jersey in 1973 and discovering there was only one telephone bridge in the state.

There is another kind of telephone conference call, which, instead of joining isolated individuals into a "telegroup," joins isolated groups together (or joins groups and individuals). Group telephone conferencing must use other means of pickup and display than the familiar individual handset.

There is a wide range of equipment known as "loudspeaking telephones," "speakerphones" plus systems that utilize regular microphones and loudspeakers available for group conferencing use. Some of these are designed and produced by the telephone companies or PT&Ts, others by variety of small manufacturing firms. With the landmark "Carterfone decision" in 1968, the U.S. Supreme Court struck down the previous AT&T rules against interconnection of their facilities with anyone else's equipment, opening the way for a wide range of independently produced auxiliary devices for the telephone. The Bell system itself originated some of the best of these to keep even with the new competition.

Teleconferencing benefited particularly from one of these — the Western Electric 50A Speakerphone. This was a tabletop device containing a small loudspeaker in a box as well as a standard handset and cradle. Two microphones could be plugged in, although only one could be used at a time. This had the great advantage of portability. Any phone outlet could take a 50A, so it could be used almost anywhere.

The University of Quebec System

A working example of a system using this equipment is the University of Quebec's teleconference system. This has been in operation since 1969 linking ten sites, seven of which are campuses of the university. The total area covered spans 800 miles in one direction and 200-300 miles in the other. Transmission is handled through the regular university switchboard with some special equipment installed to prevent accidental interruption of conferences. For the long-distance links the university leases special lines from the PT&T on a full-time basis and dedicates these to teleconferencing. Administrative and instruction purposes are served.

The University of Wisconsin Educational Telephone Network (ETN)

Probably the most extensive and active use of interactive media is the system operated by the University of Wisconsin Extension. It has been in operation, and nearly constant development, since 1965. (It is not the oldest such system, however. Albany, New York,

Medical College began a continuing education hospital network using FM radio as far back as 1955.) The purpose of the Wisconsin system is to make university-level courses available throughout the state.

To bring continuing education to all parts of the state, without at the same time putting an intolerable burden of travel on the university staff, a teleconference system was needed. Since over 200 separate locations were eventually involved, interconnected by some 5000 miles of transmission channels, video was totally out of the question financially, even if the necessary video cables had already been in place, which they were not. An audio distribution system was already in existence, however — the interconnected service of some 20 telephone companies. To begin with, the standard telephone company conference services were used, along with speakerphones. The fact that the ETN is used for instructional purposes does not in any way reduce its importance to a book on teleconferencing. It has been well established that the best teaching is interactive, and interactive teaching at a distance is one of the most important uses for teleconferencing.

Plain old telephone service (POTS), with its operator-assisted calls, was soon found to be too unreliable (even in the United States); there were too many delays in setting up calls, and too many interruptions in the midst of conferences. Consequently a network of dedicated lines was assigned to the university service on a full-time basis. Since new lines were going in anyway, four-wire transmission was installed (POTS is duplex — two-wire). In addition, small loudspeakers and then larger loudspeakers were installed, but the standard telephone handset continued to be used for response. The four-wire service provided higher-quality audio and was not susceptible to electronic feedback (a build-up of internal "white noise"), since each channel (incoming and outgoing) at each location utilized a separate pair of wires. *Acoustical* feedback is another matter, which cannot be eliminated by the use of four wires; the ETN solution to this problem is described below.

The reader will undoubtedly have experienced the high-pitched squeal or howl that is caused by acoustical feedback. This happens when sounds make a complete circuit of an audio system, returning to their point of origination, where they reenter the original pickup microphone again for another trip around the system, thus reinforcing themselves and building up to a high volume and pitch.

It is called "acoustical" because the source of the trouble lies in one or more links where the electrical signal becomes sound and travels through the air from a loudspeaker to a microphone that directly or indirectly feeds that loudspeaker (see Figure 16). It will happen when loudspeakers at one or both ends of the line are loud enough, or close enough, to the microphones that their sounds are picked up. The problem can also become acute when a loudspeaker is placed so far distant from its audience that it must be run at a high volume to be heard. Under these conditions, reverberant sound can be loud enough to be picked up even by highly directional microphones that are facing away from the loudspeaker.

The first attempt at eliminating acoustical feedback in the ETN was the simple method of cutting off the loudspeaker at a listening location whenever the telephone handset was lifted from its cradle. Of course, the user had to remember to hang it up again, or push down the cradle button, before the loudspeaker would work again. This was easy to forget, because the user was of course hearing everything on his or her handset. This sytem worked, but with additional growth and complexity of programming, further improvement was necessary.

The next step was to use the 50A conference phone, described above as a "post-Carterfone" development by AT&T. This speakerphone was specially adapted to four-wire input, but the speaker was small and there were only two microphones usable, one at a time. This seriously limited the size of audiences and required participants to pay constant attention to the switches on the device.

After considerable study and design effort, the present ETN system was developed and went into operation during 1973 and 1974. Darome, Inc., of Harvard, Illinois, was contracted to build the necessary new response equipment. This led to the development of the "Darome Convener," which is described in detail in Monson et al. (1977); here there is space to discuss it only briefly. In general it can be said that the later needs of this system encouraged development of a more sophisticated and complex system technically, which was at the same time simpler to operate and more transparent to the user. In our view, the designers of television

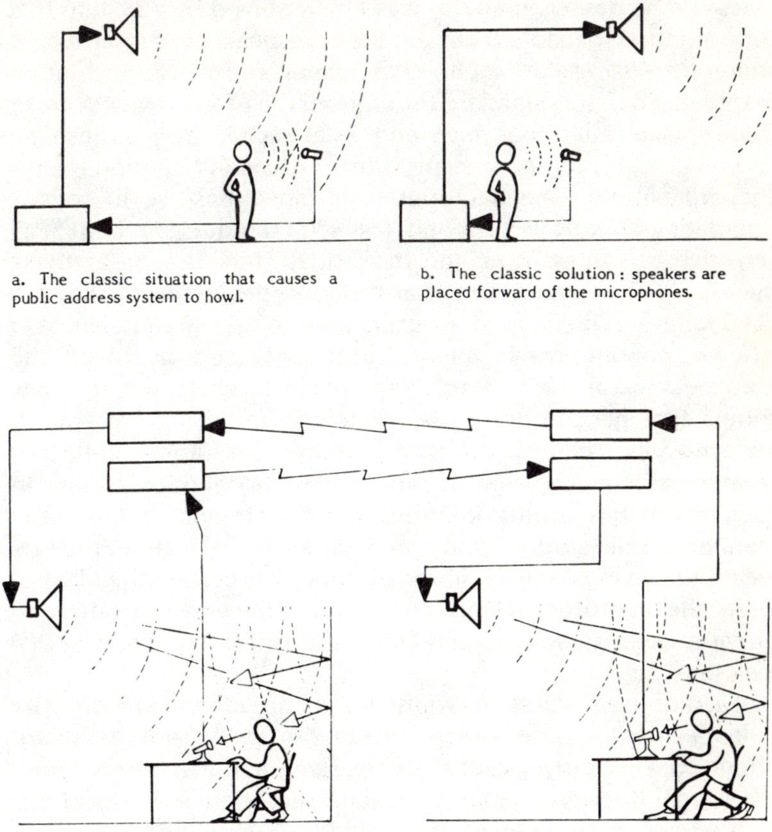

a. The classic situation that causes a public address system to howl.

b. The classic solution : speakers are placed forward of the microphones.

c. How acoustical feedback can be generated in an interactive, audio-based teleconference system.

Figure 16: Acoustical feedback.

systems have much to learn from a well-designed and thoroughly tested audio system such as this. While audio is the basic medium of video teleconferencing, most video users are so preoccupied with the visual elements that audio tends to be neglected.

The problem of acoustical feedback, which plagues all interactive systems based on audio, was finally solved by a system that *softens* rather than totally cuts off the loudspeaker in a room where someone is using a microphone for response. The Darome equipment, which is now standard throughout the ETN, provides one or more speakers and four microphones at each location. These are first of all highly directional, picking up sound in a cardiod (heart-shaped) pattern. The microphone is most sensitive to sounds originating directly in front of it, less so at the sides, and not at all sensitive to sounds from the back side. Thus as long as these microphones are not pointed at the loudspeaker when they are used (which is unlikely, since users tend to face the loudspeaker when responding, hence their microphones face away from it) the loudspeaker can be run at a reasonable level. However, many rooms that people must use for teleconferencing have hard-surface walls, ceilings, floors, or window-glass areas. Reflected sounds easily reach the front side of directional microphones in such reverberant rooms. Refurbishing of conference rooms in the manner of radio studios cannot be required, since many of the ETN locations are in hospitals, libraries, courthouses, and the like not under the control of the university. Nor can the university afford to provide carpets, drapes, and other acoustic treatments for 200 locations.

Each of the Darome microphones is equipped with a drawbar switch across the base, which must be pressed to turn the microphone on. This is to prevent a lot of extraneous noise from entering the system through open but unused microphones (200 locations with several microphones each would be a very large number of sources for such room noise as coughing, chair scraping, and the like). These switches, unlike the earlier handset cradle switches, do not turn off the loudspeakers. The loudspeaker in each room is, however, controlled in a different way by the associated microphones. If a signal arrives at a loudspeaker at the same time that someone in the room is talking on a microphone, the loudspeaker automatically becomes a "softspeaker." Similarly, if someone attempts to speak on a microphone while sounds are coming in on the loudspeaker, his or her *microphone* is automatically dampened. Thus the important "interrupt" feature, which was totally lost in the handset system, is retained; but to prevent acoustical feedback, the interrupter is put at a disadvantage and is harder to hear.

In large meeting rooms, when a person speaks into a microphone his or her voice must be amplified to be heard throughout the room. At such times the loudspeaker must function in a public-address mode. This is the classic acoustical feedback situation (see section a of Figure 16). We are all familiar with the portable PA system set up at a picnic or community event, usually by an amateur, who places a microphone at the front of the stage and loudspeakers at the rear, so the loudspeakers feed directly into the microphones, and wonders why the system squeals and howls. The softspeaking control at the ETN effectively eliminates PA feedback by automatically holding loudspeaker level below the point at which squeal can begin.

There is a range of volumes, just below the level that causes squeal, that creates sufficient reverberation in some rooms to create an echo effect, so that it sounds as though the talker were in the bottom of a barrel. On installation of this equipment in a conference site, loudspeaker volume is set by an internal screwdriver adjustment at a level just below what would cause the barrel effect. This level depends on the acoustics of the particular room. Some rooms are so bad acoustically that the loudspeaker cannot be on at all when a microphone is in use.

The Darome system has an auxiliary output (in case anyone records the conference) and an auxiliary input (in case anyone wants to feed in previously recorded material). The system can be expanded to accommodate up to sixteen microphones in one auditorium. In large classes several students share each microphone; individual microphones are possible only when groups are very small.

Darome also produces a system for use on standard POTS lines (single pair of wires). These units can accommodate up to four microphones (or more with auxiliary equipment) and can be used anywhere that telephone service is available. The same high-quality microphones and amplifying equipment are used; the main difference is that a voice-operated switch automatically reverses the direction of the line. Thus a terminal either transmits or receives, depending on where the loudest sound is coming from.

Thus all adjustments on either the two-wire or four-wire Darome equipment are fixed; and, beyond connecting the system to power, the only control is the on-off switch on the microphone. Once the equipment is installed (set in place, actually) it should give satisfactory service unless the room is rearranged, the microphones are used when they are facing the loudspeaker, or people are too close or too far from the microphone when they speak, or handle and move the microphone while it is on. In addition, a continuing effort is made to at least inform, if not train, users to recognize these misuses and avoid them.

A production center is provided for the ETN, located at the University of Wisconsin Extension headquarters in Madison. This consists of a typical radio station control room fed by three surrounding studios of various sizes. A twenty-channel console accepts the following inputs:

three studios (four to eight microphones in each studio are premixed before reaching the console)
one remote channel for originations coming from other parts of the campus
one four-wire ETN input
one two-wire POTS input with bridge controls for conferencing
two cartridge tape playback machines
four reel-to-reel tape machines, one of which is playback only
one operator's microphone in case of technical problems
seven spare channels for expansion

There are eight outputs from this control point as follows:

three studio loudspeakers
one remote line (for a loudspeaker or handset at campus location)
one output to the four-wire ETN line
one output to POTS
three outputs to reel-to-reel recording machines

An evaluation was made in 1975 after this equipment had been working for about a year. A random sample of over 600 participants rated ETN's technical equipment about 95 percent perfect (8.5 on a scale of 1 to 9).

Audio, with or without visual augmentation, is the most common teleconferencing medium in use today. In a study of some fifty educational teleconferencing systems, seventeen reported using the Darome convener, fifteen the Western Electric 50A, five other speakerphone or amplified telephone systems, and only one standard telephone conference calls (Monson et al., 1977). Hough and Panko (1977), who did not describe telephone systems of any kind, listed another nine or ten audio systems not included in the list of fifty used by Monson et al. (1977). Only five video and two computer conference systems were described in Hough and Panko's study.

Other media were available to supplement or augment all but 6 of the 38 audio systems listed by Hough and Panko (1977). It thus appears that straight audio without some augmentation is rare. In the next chapter, we will discuss some of the ways in which audio teleconferencing has been improved by visual augmentation.

3 AUGMENTED AUDIO SYSTEMS

Speaker Identification

The familiar problem with all audio systems is that since more than one person participates at the other end of the line, it must be kept clear who is talking. If the number of participants is small and there are clear distinctions between voices, such as between male and female, there may be no problem. But this is the exception. Usually a convention must be followed requiring that anyone not actually called on by name identify him- or herself whenever starting to speak. This is easily possible, but it is a habit that takes a while to learn. Meanwhile, participants are forced to direct part of their attention to the needs of the medium, instead of to the people they are speaking with and the subject they are discussing. The designers of audio teleconferencing systems have wanted, so far as possible, to make their systems transparent, so that users will be unconscious of the systems themselves. For this reason they have attempted, in many cases, to automate speaker identification.

The Canadian Department of Communication (DOC), whose developments in telecommunication have been consistently innovative, for two years operated a two-node system between Ottawa and DOC research headquarters twelve miles away. A bank of translucent panels (nameplates) was provided at each end of the line, on which could be written the names of the participants at the other end. As participants spoke, their nameplates were automatically illuminated at the other end, and they did not need to remember to announce their names. This was accomplished by providing a separate microphone for each speaker, plus a device that detected which microphone was picking up the loudest sound. A simple pulse signal was thus automatically sent to the other end, which switched on the proper nameplate.

The Remote Meeting Table System

The most sophisticated audio system in this regard is the RMT (remote meeting table) devised by the Communications Study Group for the British Civil Service. It has been a very useful, inexpensive, and effective system, and the number in use has grown yearly. It is now produced by a British manufacturer for general sale.

The RMT is a binodal system; it would be possible to extend it to three nodes, but we have found no record of its having been used this way. In the RMT system not only are distant speakers identified by lighted name plates, as in the Canadian system, but each participant is represented by a separate speaker. Thus participants are identified by the locations on the table from which their voices are heard, as well as by the lights appearing and illuminating the nameplates thereon.

Each circular meeting table in the RMT system is "set" for twelve persons, but only six chairs are provided. Every other position is a phantom place, representing a person who is present only in voice, not in body. In body that person is sitting in one of the six chairs provided at an identical twelve-place table at the far end of the line, where in turn there are six phantom places for the first six participants.

Six microphones are required at each table for the six corporeal participants, and six individual loudspeakers represent the remote group. Only one loudspeaker is activated at any one time, depending on which microphone at the other end is picking up the loudest sound. (A seventh loudspeaker is located in the center of the table; when none of the participants' individual microphones receives a sound louder than a certain level, this central loudspeaker is activated and whatever background or room sound is transmitted emanates from there.) We have been told that the illusion of sitting at a table where all twelve participants are actually present is very strong, and that one unconsciously tends to turn toward a distant person's phantom location when one has something to say to that particular person.

The entire two-way transmission of voice plus loudspeaker switching information requires no more than one or two duplex (two-way) voice grade lines at the standard cost of long-distance phone conversation.

This is one of the most successful teleconferencing systems, acceptable both to management and to its users — to management because of its low capital and operating costs, and to users (again by virtue of its low cost) because it can generally be installed where most convenient, often within the building where its users are located.

Support Materials and Complementary Media

Many, probably most, communication needs can best be served by a combination of media, integrated in software by virtue of preplanning and production. Such combinations generally concern one-way communication or the one-way presentational portions of interactive programs, so they are not strictly the concern of this book. But because presentation is often a part — the first of the three actions — of interaction, this subject deserves at least a mention.

Presentations are sometimes made by audiovisual media (such as videotape) at remote conference locations, followed by an interactive period, using in many cases regular telephone circuits. Dr. Charles Bridgeman, at Wright Air Force Base in Dayton, Ohio, pioneered in this field by teaching a course in nuclear physics to a class in Albuquerque, New Mexico, some 1500 miles distant. Dr. Bridgeman recorded his presentations on videotape, which he mailed to New Mexico, keeping a copy in his office. After he contacted his class by phone, both tapes were started simultaneously. If a question arose during the presentation, it was a simple matter to stop both machines, discuss the question, and then on cue start them both up again. While synchrony was not exact, it was close enough for the purpose.

Most audio systems, as we have previously noted, are augmented by some kind of visual means. Out of ten different audio systems described by Hough and Panko (1977), six incorporated facsimile equipment. This was used either to distribute visual materials in advance or during a conference, or as a means of permanent record of the visual materials used (to save the time required to copy charts, tables, and so on by hand).

Printed Materials

Because the preparation of printed and pictorial materials takes time, they are usually used only when a teleconference is planned well in advance. Routine business meetings, whether teleconference or face-to-face, rarely go much further than to circulate a one-page agenda among the participants at the start of the meeting, unless perhaps a document is the subject of the meeting.

In instructional teleconferencing, however, planning can often start months in advance. Instructors in University of Wisconsin Extension courses distribute printed outlines of each subject they will present and discuss in conference. This has several instructional advantages: It cuts down on note taking, which often reduces listening because students are too busy writing; it helps students structure the information in their minds; it provides a ready means to review a presentation; and it provides students that have a visual, print-oriented learning style with a second chance to comprehend what may have been difficult for them to grasp when spoken.

In cases where text materials for a course would be difficult for students to obtain, full syllabi are sometimes written for a telephone course. These may contain equations, diagrams, or pictorial materials, depending on the needs of the course. They may be referred to during a lecture, as, for example, when a medical discussion is based on an anatomical diagram or a photographic example. The fact that such graphic aids are recorded in hard-copy form means that the instructor cannot point things out spontaneously, nor can students indicate specific details in questioning the instructor. Thoroughly labeled diagrams can partially overcome this limitation, but there is not too much one can do in labeling a photograph. Also, printed materials are generally limited to monochrome; color reproduction in page form is either very expensive or limited in realism.

Projected Materials

When color pictures are important, the University of Wisconsin Extension teachers often use 35 mm slides. This, too, can be expensive. If a course has sections meeting simultaneously in thirty hospitals, thirty sets of slides must be made and distributed in advance. Either a projectionist must be found at each reception point and paid by someone or (as often happens) one of the enrollees must take over the responsibility. The slides are numbered and the instructor can cue slide changes by number.

Another alternative is to use overhead transparencies. The instructor can usually make his or her own original "artwork" and run off transparency copies on an electrostatic machine such as Xerox or a heat-sensitive copier such as Thermofax. Since overhead projectors are as common as slide projectors in most learning locations in Wisconsin, this alternative takes less time than slide making and is less expensive. Color is not possible in locally produced transparencies (except the kind where ink is lifted off a magazine page), although some color may be added by hand, such as underlinings for emphasis.

The NASA and Boeing Audiovisual Systems

Some of the biggest teleconferencing systems have been based on audio but augmented by visual materials transmitted in advance by facsimile systems and projected on large screens in conference rooms in several locations. Among the earliest of these were the systems developed by the U.S. National Aeronautics and Space Administration (NASA) and the Boeing Company in support of the Apollo "Man on the Moon" project. The conference facilities of NASA and Boeing are compatible, and frequent joint teleconferences are held. Within Boeing alone it is reported that as many as 250 people have taken part in a single teleconference. Audio was originally switched manually, with request-to-speak signals transmitted to a central controller. More recently, voice-actuated automatic equipment has replaced the manual switcher.

The network is operated by NASA through a central point of control in Huntsville, Alabama; any point may be immediately connected to any other single point simply by dialing the number. Four-wire circuits are used for high quality. When more than two locations are involved in a conference, a conference operator is required to set up the necessary links.

Each of the conference rooms in these systems is equipped with three built-in rear-projection screens, making a total projection area some fifteen feet wide by about four feet high across the front of the room. Visual materials to be used in a conference are prepared at one of the centers and transmitted to all of the others by facsimile. The Xerox LDX machine was used for the Apollo project, sending at a rate of forty seconds per page. This required a fairly expensive wide-band transmission — equal to twelve voice channels. At each location the LDX receiver produced a Xerox copy. At first these Xerox prints were projected directly via an opaque projector (Epidioscope). Later it was found better to go through an additional step using the Thermofax machine to convert the opaque Xerox pages into transparencies. Overhead transparency projectors put more light onto the screen and produced a clearer image; they soon became standard equipment behind the NASA and Boeing rear-projection screens.

During the Apollo project, visual materials to be used for the presentation portion of a conference were sent out shortly in advance of the conference, carefully numbered both for sequence and for screen. If during the ensuing discussion additional visual materials were needed, from any source, these could be called for, transmitted to all locations, and projected in about two minutes, handled by technicians in adjoining rooms.

At the end of the Apollo project the system was to a large extent retained, but a slower-speed facsimile transmission system was substituted. The system was then used for other NASA projects, such as Skylab, Voyager, and Space Shuttle. By the end of 1976 there were some thirty conference rooms in various parts of the country connected to the NASA system, and more were being added.

Facsimile

Facsimile (or fax) is a means of transmitting prerecorded materials; is it not a live medium. No one can originate a message on a facsimile machine. It might be possible for two people, each with transceivers or separate transmit and receive units, to communicate back and forth by typing up or writing out each message then sending it by facsimile, but it would be a very cumbersome method. Also, low-cost fax machines tend to be slow on transmission. Some expensive machines will transmit a page in twenty to thirty seconds, but most require several minutes. For use in teleconferencing, the usual role of facsimile is ancillary. It is used to augment the AT&T Picturephone Meeting Service (see Chapter 1), and its uses in the NASA and Boeing systems were described above.

Writing, Drawing, and Pointing

A large number of telecommunication applications require no more than alphameric and diagrammatic means; the full gray scale of the photograph or the high definition of facsimile is not required. In many of these uses — engineering courses, to give one example — these limited visual means are not merely desirable, they are indispensable. Mathematical formulae and technical diagrams can not be conveyed by spoken words alone, except in the case of necessity, and then only with the greatest difficulty. Untrained people have done emergency operations at sea when there were no doctors available, by contacting a doctor and following directions by interactive radio. No one would attempt to teach physiology, however, by nonvisual means, and neither could a math instructor operate without a blackboard or its equivalent.

In an earlier volume, Bretz introduced the term "semimotion" to cover the "motion" of writing, drawing, and pointing. Each of these motions is very simple: the gradual extension of a line in the first two cases, the movement of a spot or pair of cross hairs in the third. Where the full motion provided by film and video must record or transmit the myriad details of a full picture 24 to 30 times in the space of a second, semimotion need only reproduce a tiny segment of a line or the slight displacement of a pointer. Thus it can be transmitted (while the line is actually being drawn, or the pointing spot manipulated) on a channel of very limited bandwidth —about one-fourth that of a standard voice-grade line; something like one-five-thousandth of the bandwidth required for motion video.

A familiar example of semimotion is telewriting. The earliest device for transmitting handwriting at a distance was the "telautograph" invented by Elisha Gray, the inventor who had developed the telephone and applied for a patent just a half hour after an application had been received from another inventor, a Mr. Alexander Graham Bell, for the same kind of device. Except for that half hour we might today have the Gray telephone system in America instead of the Bell system.

The telautograph (now a generic term) worked by constantly transmitting the movement of a stylus, which was used for writing the message. The surface written upon was unimportant. The stylus was held in a kind of two-dimensional harness, which moved mechanically with the stylus and sensed where the stylus was placed and in which direction it was moved. At the other end of the channel (or ends, as the case might be) a similar stylus moved a writing pen and simultaneously traced the same lines.

Telecommunication of handwriting is superior to voice communication in that it leaves a permanent record, while voice is evanescent and disappears as it is heard. It is inferior in that it takes about five or six times as long to write something as it does to say it. Used as a complement to audio intercommunication, however, telewriting provides a means to send simple graphics, as one might draw on a blackboard, while any expository narration is done by voice.

Telautograph has been widely used for intercommunication within industrial plants, such as in the transmission of orders from one department to another — as in railroading, for instance, from one location on the line to another. It has been estimated that there may be some 25,000 transmit or receive telautograph units installed in America today. Three manufacturers produce and market the equipment.

Eventually, equipment was developed with which telewriting could be displayed on a large screen for group viewing, and this lent itself to instructional purposes. This was done by reflected light, in the manner of an opaque projector, or by transmitted light, using an adapted overhead transparency projector. These were phone-coupled; after the connection for such a transmission was made by phone, the handset was laid in a cradle on the display machine and the transmission could proceed. The audio portion of this media combination was transmitted over a second phone line, which terminated at the receiving end in a conference phone.

Nearly all applications of telewriting were one-way. Very few were interactive, although many combined the medium with audio as described above, and the audio provided an interactive component.

Victor Comptometer, in their Electrowriter (which is no longer on the market in the form here described), provided a means for incorporating local graphics as a further component of what then became a combination of three media: audio, telewriting, and local graphics. The Electrowriter projector was of the overhead transparency type. The local graphic was a transparency that was sent out to places of reception in advance and placed in the projector under the moving pen and transparent sheet on which it was writing. Thus a map, for instance, could be placed beneath the writing stylus. With identical maps on all receiving machines, the telewriter operator would be able to point to locations on the map, or draw weather information, for example, upon it.

The Electronic Blackboard

The Bell Telephone System developed, during the 1970s, a semimotion transmission system that they called the Electronic Blackboard. In 1974 it was tested and evaluated at the University of Illinois, and by 1978 it was offered by AT&T as a regular telephone company service. The rationale behind the Electronic Blackboard was the transparency principle: The most easily used system is the one that is least noticeable to the user and requires the least acquïsition of special skills or habits.

Teachers of math-based subjects, such as engineering, for example, are accustomed to depending on the classroom chalkboard, as indeed are teachers of many other subjects. Some cannot feel comfortable talking unless they have a piece of chalk in hand.

AT&T developed a board, outwardly like an ordinary blackboard, a little larger than three by four feet, that could sense the marks that were made on it and transpose these to electrical pulses (see Figure 17). These in turn could be transmitted on a standard voice-grade telephone line. At the receiving end the original marks were displayed on a standard TV monitor. The user could apply his or her regular blackboard skills and techniques, relating in the standard way to a face-to-face group. The only unfamiliar thing in the new teaching/speaking environment was the presence of a TV monitor. The only new habit to be learned was that of checking this monitor from time to time for assurance that the writing was properly reproduced and large enough to be clearly legible.

This black-and-white TV monitor is easiest for the blackboard user to see if it is placed alongside the board. This, however, puts it much closer to the user than any viewing screen is to the participants in the distant location(s). The maximum distance that people in a small group tend to sit from a television set is about 10 or 12 screen widths. Thus if the speaker's monitor is placed 12 screen widths away from him or her, the perspective will be similar to that of his or her furthest viewers, and the speaker will get a far more realistic idea of the legibility of the characters he or she is writing. If the speaker is using a 14-inch-wide monitor, this distance will amount to 14 feet. If the monitor screen is 22 inches across, it should be placed 22 feet from the speaker. Of course, in some applications, groups may be smaller and the maximum viewing distance may be 6 screen widths in each case. The relatively small size of the television screen is something that is not realized by most of us — we are used to the small television set in the home (where we tend to sit not more than 6 screen widths away). If one were to invite one's friends in and proudly present vacation films or

NOTE: The underline being drawn on the board is appearing on the monitor at the side. Photo reproduced with permission of AT&T.

Figure 17: The Electronic Blackboard.

slides on a screen less than two feet wide, they would laugh. Yet there is equipment on the market that would make it possible to display the same films or slides on a TV screen, which is even smaller than 2 feet wide, and no one would laugh at that.

If the Electronic Blackboard system is to transmit both voice and writing/drawing simultaneously, as in instruction, two duplex phone lines are needed. If voice and writing can alternate, a single duplex line can suffice. To free the speaker from an attached microphone and cord, Bell Telephone recommends the use of a conference phone for the audio portion of a presentation, although better-quality sound can of course result from use of a closer microphone.

Once a board has been filled, it can be cleared by pressing a single button located at the board. That is, the image is cleared from the TV sets; the chalk must still be cleared from the board by an eraser. If only a word, a single character, or part of a diagram is to be erased, this can be done in the normal way using the standard blackboard eraser. When the eraser is not in use, it rests in a small tray centered below the board, where its presence is automatically detected. If it is taken off the tray, the system switches immediately to erase mode, and whenever the board is touched, the marks at that point will disappear. It is only pressure, of course, that operates the mechanism; the board may be written on, or erased, with the finger. Teachers who are accustomed to making a small erasure with a finger have to remember to lift the eraser out of its tray and put the system into erase mode, or the finger will add rather than subtract.

While one-way blackboard and two-way audio is the usual combination of media for interactive communication with the Electronic Blackboard, full visual interaction is also possible. If a board is provided at two locations, for example, both boards may be used to write or draw on the same visual space, and the common result of both originations will appear on all screens. The entirety of both boards will be erased by pressing the "clear" button at either one, and if either communicant lifts his or her eraser from its tray, the whole system is put into erase mode and neither person can write.

In visual interaction, therefore, there are some new procedures and habits to be formed before the system can become transparent. The TV monitor must be watched closely when adding or subtracting information that has been put there from another terminal. However, when two phone lines are being used and there is constant audio interaction, sufficient communication is possible to quickly right most problem situations. When three or more locations are interacting in a conference call, for instance, and it is noticed that writing is not appearing on the monitor, the chairperson can say, "Check your monitors and erase anything from your boards that is not reproduced on the TV monitor," and then, "Make sure all erasers are in their trays."

An important feature of the Electronic Blackboard is that whatever is written on it can be simultaneously recorded on an audiocassette. A standard stereo tape recorder is used, one track recording audio while the other records the semimotion visual component. Thus the entire series of events that took place during the period of recording can be exactly reproduced on playback. Of course, the board itself has no function during playback; all that is required is a TV monitor and its auxiliary electronic equipment, plus the stereo tape recorder.

During the testing phase of this system in 1974, an advanced course in engineering was taught at the University of Illinois in Urbana to students at two distant locations, the Sunstrand Aviation Corporation at Rockford, Illinois, and the Caterpillar tractor plant at Mossville, Illinois. This particular application was notable in that three Electronic Blackboards were used, side by side, and each participating location used three monitors. As one board was filled, the instructor went to the next. This had the advantage of simultaneous display of a large amount of mathematical data and diagramed information so it could be referred back to at any time, even by an individual student following his or her own train of thought. This system required two duplex lines, for audio and the first board, and two simplex lines (one-way) for the other two boards. There was no visual interaction; only the audio was interactive. At the time of this writing, this triple system was still in operation on a monthly lease basis.

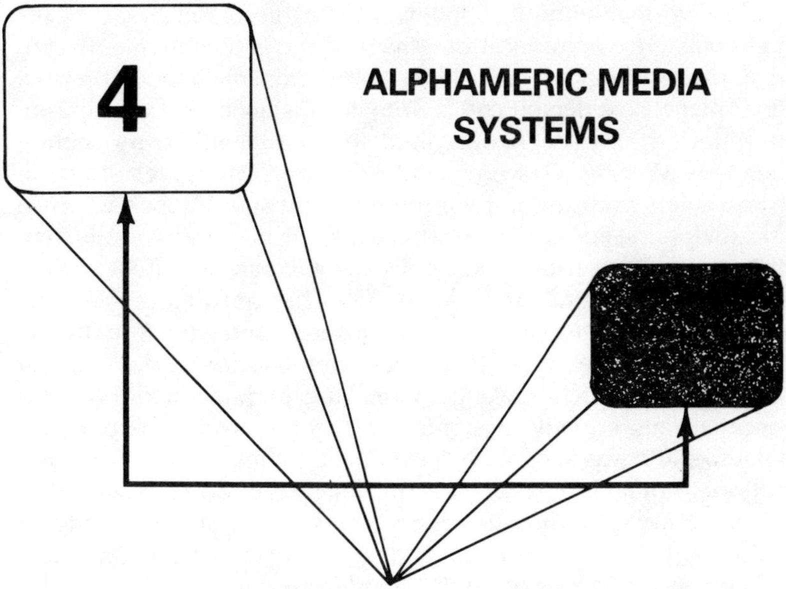

4 ALPHAMERIC MEDIA SYSTEMS

The use of the voice in interactive communication is so universal and predominant that we tend to forget there is another means of verbal communication that also is useful for interaction. Writing was first developed as a means of recording the spoken word and making it permanent, and has thus generally been a one-way medium. Personal two-way correspondence can be interactive — when each correspondent responds to the other — but the time between a message and its response is generally so long that letter writing is a far cry from face-to-face conversation. In recent years, however, a new alphameric means has emerged that, when used to communicate between individuals, is called "electronic mail." This consisted originally of messages computer users sent to each other, which were stored temporarily in computer files. Today under the name "computer conferencing" the medium has expanded to include teleconferences involving dozens of participants nationally or internationally distributed. It has the potential of developing into a widely used communication means intermediate between the telephone and correspondence. It has also proved practical to type back and forth in a manner very close to spontaneous conversation.

Telewriting (without sound) was the earliest alphameric means of interactive communication. An invention of the nineteenth century, the telautograph made it possible to transmit handwriting as it took place, so a person could write at a distance. The workings of this device have been described in Chapter 3, in the section headed "Writing, Drawing, and Pointing." Using telautograph transceivers (transmitter-receivers), two persons at opposite ends of a telephone line could write back and forth interactively. In fact, they could both use the same visual space and in effect correct each other's writing. To a person who has not before seen this device in action, it is as though an unseen hand were forming the letters at the receiving end. We have noted before, however, that writing is very much slower than speaking, and since audio systems were always equally available, and generally less expensive, telautograph was very rarely if ever used for interactive communication. Its main use has been to transmit orders on a one-way basis within an organization, such as a railroad, where it is important that a permanent record of commands be retained for accountability and for ready review to avoid misunderstanding.

Another early means of alphameric communication was the teletype machine, a typewriter that transmitted characters as they were typed, to appear on distant typewriters operating in receive mode. Here again, real-time interaction was a possibility, but was rare in actual application. There are still some situations, however, in which teletype is far superior to the telephone in several particulars. The fact that teletype produces hard copy means that the message may be retained, studied, and reviewed, all of which are in contrast to the evanescent nature of the spoken word. To record a telephone call requires a tape recorder with a special phone pickup and legal or informal permission generally, and then the resultant recording must be transcribed into typewritten form before it is convenient for reference and retrieval.

It is interesting to realize that the famous "hot line" between Washington and Moscow, which has the vital function of providing immediate interactive communication in case of international incidents that might be misinterpreted as warlike acts, is not an audio telephone as most people imagine. It is a telex system, relying on the printed rather than the spoken word. This assures better understanding and immediately produces a recorded message (when minutes count) that can be more readily reproduced and studied than a message in the audio medium.

We have had recent experience with international calls to and from Pakistan, where overseas calls are handled manually. Telephone calls to and from the United States cost about $4.00 per minute and must be booked generally three or four days in advance. A live teletype connection, on the other hand, costs less, and may be made within minutes without operator assistance if the distant party is available. Although this is a relatively rare use of the telex service, a live interactive conversation may be held via this medium. Whatever each party types will appear on both printouts. Of course, the two cannot type simultaneously at both ends of the line or the words and individual letters of one message will overlap the other and no sense will result. Thus the familiar "over and out" sort of convention, common when two parties are interacting on a single radio frequency, must be followed. It's not the same as hearing a person's voice and being able to grunt or "uh-huh" whenever you want, but it is certainly better than waiting four days, and it leaves you with a printout.

Keyboards

Alphameric communication requires a keyboard for the sender to encode a message by typing it letter by letter. It would seem that ability to type might affect the usefulness of this medium, but, according to several studies, this does not necessarily seem to be the case. Evidently, the time that a user spends thinking up what to type next will in most cases far exceed the time it takes to type it, whether the user is a touch typist or the hunt-and-peck kind.

Of course there is quite a range of typing speeds in each of these categories; a good hunt-and-peck typist can exceed the speed of a slow touch typist. A slow hunt-and-peck typist, on the other hand (sometimes referred to as the "hunt and hunt and peck" variety) can be very slow indeed. People who are totally unfamiliar with the standard typewriter keyboard (called the QWERTY, for the letters on the first six keys) will find a keyboard with a straight alphabetical arrangement (the ABCD type) far easier to use. Everyone is already familiar with each letter's alphabetical order. Reuters uses the ABCD keyboard for its quasi-interactive news and stock reports service, described in the next chapter.

It is not always necessary to have a keyboard with a separate key for each letter, accent, numerical, and punctuation mark. Mode-changing keys are generally used to allow the other keys to do double or triple duty. The familiar shift control on the standard typewriter is such a control key. When the shift key is depressed, letter keys type upper case and numeral keys type symbols such as % and #. Computer keyboards have another such key (generally called "control"), which produces yet another set of functions for some keys. For example, in using the interactive text editor called XED, you must type two keys, control and Z, to put the system in the command mode. Then the R key is pressed if you want to read or the W key if you want to write. Typing a (for append) or i (for insert) puts you in text entry mode, after which typing Z produces a Z, typing a produces an a, and so forth. Thus the R key, for example, is used for three purposes: R, r, and read.

The idea of using one symbol for two meanings preceded by a special symbol to indicate which meaning to take is very old in communication. The old navy semaphore system, with which a sailor signaled the letters of a message by various positions of two flags, is a case in point. In designing this system it was undoubtedly felt that 26 letters and 10 numerals, plus a signal for break between words and an error signal, 38 signals in all, was too much for a sailor to learn. The number of signals was reduced by one-fourth by using the signals for the first 10 letters for the numerals as well, if preceded by a special symbol signifying "numerals follow." This was analogous to the typewriter shift or the modern computer control key.

There are ways of encoding all the letters of the alphabet on keyboards that have a very limited number of keys. These are not intended for producing text, however, but for such purposes as information retrieval — obtaining stock quotations, for example. The standard telephone touch-tone pad or rotating dial carries 24 letters (all letters except Q and Z), with 3 letters sharing the position of most numerals (the numeral 1 is alone). However, the telephone normally does not distinguish among the three letters on the typical key. It has no shift key. The letters are left over from a day when it was considered easier for subscribers to remember and dial letter-numeral combinations than numerals alone. The telephone itself knows only numerals.

When a computer is at the other end of the line, however, it can be programmed to "shift" to letters when, say, the star key on the touch-tone pad is pressed. Then it is only a question of which of the three letters is meant when a button is pressed. This is easily defined by pressing a second button, a 1, a 2, or a 3, to indicate the first, second, or third letter that shares the key. Q, Z, and the dash (minus sign) are added to the 1 button. Thus pressing a 6 and then a 3 would signal the letter O, since O is the third letter that shares the 6 key. A, B, and C share the 2 button, so if one pressed 2 and 2, the computer would understand B (see James, 1971: 149).

Any of these keyboards, either full or abridged, must be learned. They also, of course, require literacy. Alphameric communication requires first that the users be able to read and write. While illiterates are able to use the telephone, radio, and television without any problems, they are helpless when confronted with printed words. The ability to spell is also an important requirement.

But literates include fast readers and slow, fast touch typists and slow hunt-and-peck typists. It is only the touch typist, with years of experience at the QWERTY keyboard, to whom the alphameric media can possibly be considered transparent. Most accountants, and bookkeepers, and supermarket checkers are skilled one-hand typists on numeric-only keyboards, but may be as helpless as most of the rest of us when it comes to full keyboard typing. Abridged alphameric keyboards are used so little, however, that most of their users remain at the hunt-and-peck level.

Originally, all keyboards were permanently attached to computer terminals or teletype transmitters. In recent years, however, portable keyboards have been developed, which can be carried in a briefcase and used to transmit alphameric information via phone lines. Because most telephone systems are still analog (transmitting electrical signals analogous to the sound frequencies they represent), the digital output of a keyboard must be converted to sound frequencies for transmission. For this purpose the "acoustic coupler" is used. It may be a separate unit or built into the body of a portable keyboard. The computer is contacted by ordinary phone, and when the computer tells the user (by transmitting a "carrier signal" tone) that it is ready, the ordinary handset being used is placed into the acoustic coupler cradle. The coupler transduces the digital pulses into sets of audio tones that are transmitted by the phone, recognized by the coupler at the other end, and transduced back into digital pulses. In future, when all telephone systems have gone digital, this double translation process will not be necessary.

Real-Time Interaction

Like the interactive use of telex, previously described, computer users can type back and forth in real time. In the TENEX series of computer systems, for instance, this mode of operation is called Link. Four people may be linked together at one time. This system is particularly useful for two people who are working on the same project but are too far apart to meet regularly face to face. Even if the distance between them is measured in thousands of kilometers, linking for extended periods of time can be inexpensive, relative to telephone costs, for example.

In Link, as in interactive telex, people who wish to contact each other generally go to their terminals at a mutually arranged time. Unscheduled linking is also possible; a call is initiated by one person but can be completed only if the other happens to be using a terminal at the time. The demand is not great enough to justify a ringing system such as that for telephones to call people to a terminal that is not in use. (The computer will know what terminal a person is using because users must log in by name.)

A computer user may of course be in the midst of something he cannot drop when he is informed (by the sudden appearance of a line of print on his screen) that his colleague Johnson wants to link with him. He may either refuse the link flatly, making a mental note to contact Johnson when he has time, or he may leave his work suspended, go into Link mode and engage in a dialogue with

Johnson, after which he can have the computer bring him back to the exact point where he was working before he left. As in telex, everything typed by both parties appears on both screens (or typed print-outs, if such terminals are in use). If they both type at once only a jumble of letters results, so they must take turns.

Another real-time system was developed in association with the Plato system at the University of Illinois, sometime in the early 1970s. This was called Talkomatic. The screen was divided horizontally into five segments for five participants. Others could read and watch, but could not participate until someone dropped out. Several such conferences could be going on at once, on different subjects, each dividing a full screen among five participants. It was an anonymous system, and was used for recreational rather than working purposes. One Talkomatic channel, for instance, was generally monopolized by science fiction buffs. There was no recording of anything. Each person had enough space for four or five lines. When another line was added at the bottom, each previous line moved up one and the top line disappeared.

One of the earliest real-time systems was Party Line, developed by Murray Turoff around 1970 for the U.S. Office of Emergency Preparedness (see Hough and Panko, 1977: 63). This system was applicable to a larger group of participants than Link or Talkomatic. Instead of all terminals receiving at once, simultaneously with the typing of each message, a short-term recording system was interposed between the participants, indexed and managed by the computer. All participants could be typing at the same time, or all reading at once, or any combination in between. The computer recorded all entries in chronological order, kept track of which entries each participant had read, and told participants how many there were that they had not read. If someone was late to the conference, the computer fed him or her a transcript of the entire discussion that had transpired before his or her arrival. There was also a way to recall a previous entry in order to reread it. In this system everyone was asynchronous with the system; no one read material at the same moment that the sender was typing it out.

There is no facility for correcting or editing anything in these real-time systems; just as in voice systems, accuracy is traded for immediacy. In systems like Party Line, in which a number of participants are communicating at the same time, more than one thread or topic may be going on simultaneously. The conference remains unified, in that everyone receives all entries, but a response is often separated by one or more entries from the statement that stimulated it. This is not necessarily confusing, but some find it helpful to begin each entry by a reference to the number of the earlier entry, if any, to which it is a reply or comment.

If everyone is asynchronous to the system, why is Party Line classified here as a real-time system? How is it possible for something to be both asynchronous and real time? After all, in Party Line all participants must be at their terminals at an appointed time and all participate at once. Jim Carlstedt of Information Sciences Institute has defined a real-time interchange as one in which responses feed back rapidly enough that no participant has time to change his or her attention to other matters, but continues to give full attention to the conference. According to this meaning, real time is a broader concept than synchronous and a real-time interchange can also be asynchronous.

To summarize, then, in a real-time conference that is synchronous, all participants "meet" at their separate terminals at the same time. Words typed on one terminal appear on others as they are being typed. In a real-time conference that is *a*synchronous, all participants meet at the same time, but, as in Party Line, they may type when they wish, and read what others have typed when they are ready. In a fully asynchronous conference, two or more participants may "attend" simultaneously, but this is unintended and unnecessary. Each is free to come and go from the conference as he or she finds most convenient, and read or type as he or she wishes.

Real-time intercommunication, similar to live radio and television, crosses barriers of distance (or other barriers of a spatial nature, such as prison walls), bringing widely separated people into communication who cannot come together to meet face to face. Asynchronous intercommunication, familiar to us in recording media such as audio- or videotape, does all that, and something else, too. Barriers of time are also bridged. People who cannot get together at the same time because their schedules conflict may confer asynchronously. Asynchronous teleconferencing can bring people together across space *and* time. Although other recording media are capable of this, few have been used in this way. Examples are written correspondence and tape "letters"; others may appear in the future. The alphameric media, however, are more easily used for fully asynchronous interaction, especially when assisted and managed by that most powerful and versatile tool, the computer.

Asynchronous Systems

When a computer conference is separated entirely from real time, so that participants do not have to assemble all at the same time, a much more convenient system for serious use emerges. One of the earliest of the many asynchronous computer conferencing systems was Forum, developed at the Institute for the Future. Probably more research into the usefulness of the various features of computer conferencing systems has been done on Forum than on any other system.

In an asynchronous conference, participants devote to the conference whatever periods in the day or night seem most convenient, or, if they do not have terminals of their own, at times when they can schedule the use of one. When participants "log in" (type their names and the secret code words that give them access to the conference), the computer first informs them how many entries have accumulated since they last participated. They will want to begin by reading these, which may amount to a considerable backlog if they have not participated for several days. Having done this, participants may wish to respond immediately to some comments and think a little about others. They are under no pressure to start typing, or to keep up a steady flow of words because someone is reading it as it appears. They correct typos as they go, cancelling out letters, words, or whole lines and typing over. They do not actually enter what they type until it is complete and satisfactory. If a participant is not satisfied, he or she may cancel out what has been done and start again, or put it in a separate file, try another expression of the same thought, and then choose the one he or she likes best. On some systems there is an editing capability that allows participants to go back and take out words or paragraphs, adding a sentence here or a word there.

Such meetings extend for long periods, sometimes lasting weeks or even months, although under pressure a computer conference can be completed in a few days. Urgent meetings and meetings held to prepare for scheduled events can better be held via some synchronous medium, such as an audio system. Routine business meetings, simple information exchange, and research working sessions adapt very well to computer conferencing. Users feel that much more can be accomplished than via telephone, and, where distances are great, at considerably less cost.

To acquire some data on the characteristics of asynchronous computer conferences, researchers at the Institute for the Future (IFTF) applied several kinds of analysis and measurement techniques to ten Forum conferences. These conferences were conducted by such organizations as the U.S. Geological Survey, the Washington State Planning Office, IFTF and Bell Canada, IFTF and the Information Sciences Institute, and research collaboration within IFTF. These conferences took place during 1973 and 1974 and were probably quite representative of the best of computer teleconferencing at the time.

Table 1 summarizes all ten conferences in terms of overall length, average time devoted by each participant, number of participants, number of entries, and length of entries. The low and high ends of the range for each of these parameters is shown, as well as the mean value for each.

It can be seen that asynchronous teleconferencing is not a good way to get things done in a hurry. These ten conferences lasted from a little over 2 weeks to a maximum of 5.5 months. The average participant devoted 9.5 hours to a given conference, over an average period of 2.5 months. This amounts to less than an hour a week. It can safely be assumed that this time was not taken at the sacrifice of anything else of importance, so practically all of the communication that resulted was probably extra communication that would not otherwise have taken place. The average number of entries was 564 at 61 words each, 17,500 words being exchanged during the average conference. This number of words would fill over 50 pages of the standard 8½" by 11" professional report.

To make a comparison with written correspondence, we can look at the last column of the table and take this list of mean values as a single hypothetical conference involving 13 people, lasting 2.5

TABLE 1 Summary of Ten Computer Conferences: Field Tests of Forum, Conducted by IFTF

Parameters[a] *(per conference)*	*range* *Low*	*High*	*Average*
Number of participants	3	31	13
Length (in days or months)	18 days	5.5 months	2.5 months
Average number of hours per user	4	20	9.4
Number of entries	53	1341	564
Words per entry:[b]			
public[c]	50	160	61
private	50	160	38

a. From Vallee et al. (1974b: 36).
b. From Vallee et al. (1974: 55).
c. Public entries go to all participants; private entries go only to those addressed.

months, and so on. The number of entries per participant has not been included in these data because the spread within a conference is generally very wide and very uneven. In a typical conference reported by IFTF, 20 percent of the participants were responsible for 80 percent of the communicating. This is probably about what happens in face-to-face conferencing also.

Translated into postal letters, at an arbitrary rate of 2 entries per letter (120 words), the total contents of this conference would require 140 different letters. If any group should actually try correspondence conferencing (which, for very good reason, has not been done), they would undoubtedly find a very different type of communication resulting — the various parameters would change considerably from those of computer conferencing. Thus the following comparison has only illustrative value.

Allowing three days for delivery of each letter and two days in which to respond, this hypothetical conference would last (if anyone were still interested) for almost four years. In considering postage it must be remembered that copies of each letter would have to be sent to twelve fellow participants, so the post office would be carrying 3384 letters before the conference was over. (Double all these figures if you think a separate letter would be needed for each entry, in order to retain the same degree of interaction as the computer conference.)

One further point in this comparison: Correspondence conferencing would be hopelessly cumbersome because of the time required for transport. With overnight delivery of mail to any part of the country, and a commitment by all participants to immediate response, transport time plus response time could be cut to a day or two. Even these heroic measures, however, would not shorten the average 564-entry conference to less than a year. On the other hand, computer conferencing, despite its immediate transmission, might be less efficient than written correspondence where computer terminals are in short supply or have to be rented.

Computer conferencing may also be compared with spoken language intercommunication via the telephone (or some other audio system). These media are alike in that they are transmitted instantaneously. They differ greatly, however, in the time they require for the expression of communication. The only measurable parameter we have is words per minute (or characters or bits, if you like), which is of limited value since oral speech tends to be much more wordy than written text for the same thought content. One of the IFTF transcript excerpts contains 730 words in 13 entries (see Vallee et al., 1974a: 52-55). This synchronous interchange took just about 60 minutes. Communication proceeded at the rate of 12 words (approximately 1 line of type) per minute. Another example of a synchronous exchange is reproduced in Vallee et al. (1974b: 85-90). This contains approximately the same number of words, in 20 entries, but took place in only about 34 minutes. This is almost twice as fast as the first example: 21 words a minute.

Had these same interchanges taken place via telephone, they would certainly have been more wordy, but even so they would have been completed in much less time. At a talking speed of 175 words per minute the same number of words could be transmitted in a little over 4 minutes. If the number of words is to be the measure, 14 times as much could be transmitted via phone in the 60 minutes of the first interchange; 8 times as much in the 34 minutes of the second example.[1]

One of the most important features of asynchronous teleconferencing is that many threads of discussion may be interwoven into the same conference. Entry 201, to take an arbitrary example, may have no relationship to entry 200, which indeed may have been entered so recently that the writer of 201 would not have had a chance to read it. Entry 201 may be in response to entry 195, for example, made some days before. Entry 202 by some other participant may be on a quite different topic, made in response to entry 198, and so forth. Thus a computer conference started on a clearly defined topic can subdivide and branch out into a series of subconferences in which individuals quickly find others whose interests and expertise in specific aspects of the subject parallel their own, and thus all can penetrate more deeply, contribute more to a conference, and benefit more from it.

In face-to-face conferences, on the other hand, no more than one aspect of a subject may be pursued at a time, and all participants must be included. This encourages an emphasis on aspects of the most general interest; a chairperson hesitates to spend too much time on topics that involve only a small number of the attendees.

Another important feature of computer teleconferencing, synchronous or asynchronous, is that everything is recorded, indexed by entry number, date, time of day, and participant, and, if the participant provides them, by a few key words as well. No one gets left out of a computer conference unless he or she truly has nothing to say. No one remains silent because he or she is too shy to speak up, letting more aggressive people monopolize the floor. There is always enough time to get everything said that anyone wants to say. The worst that can happen to any participant is that he or she makes a contribution that the others read but make no comment upon. Assuming that all participants are seriously interested in exchanging information or solving a problem, probably more can be done with less effort in computer conferencing than in any other medium, although it may take longer. If the meeting is a confrontation, however, or if its purpose is persuasion, a video conference or a face-to-face meeting, perhaps even an audio medium with the humanizing effect of the voice, may achieve the objectives more readily.

Synchronous communication between different parts of the world has serious limitations when people in different time zones try to find a common time convenient to all. Here is where an asynchronous system can be useful indeed. Even between the east and west coasts of North America, where there is a spread of four times zones (three hours difference coast to coast), usual working hours overlap for only three hours each day. In London the business day is just closing as offices in Los Angeles and San Francisco start taking down the shutters. Either the Californians must come to work early or the Londoners must work late if they are to reach each other in real time.

Asynchronous messages, however, sent in the late afternoon from London can be received at the start of the same working day in Los Angeles and replied to any time during the Los Angeles business day. By 5:00 p.m. in Los Angeles it is 1:00 the following morning in London, still eight hours before the starting of the working day, during which the Los Angeles messages may be sent to London and be waiting for the London staff to receive the moment they reach the office. Of course, 9 a.m. in London is still only 1:00 a.m. in Los Angeles; the Londoners, like the Los Angeles staff, have a full working day to get together and formulate a reply and still have it there for the opening of business in Los Angeles.

For a continuous interactive conference the pace cannot be faster than one entry (or series of entries) each day in each direction. If the need is simply for information, however, a request can be sent from either end at the close of the business day and the reply received at the start of the next, resulting in no wasted business time whatever.

There are noticeable differences in the styles of writing that result from these various modes and media of interaction. Typing back and forth in Link, for example, tends to produce a very informal and at the same time a very wordy kind of writing. There isn't time to condense thoughts into a minimum of words; thoughts come out however they are first conceived, much more in the manner of oral expression, with occasional fragmentary sentences and incorrect grammar. Typographical errors are left where they fall. Because writing is so much slower than reading, even a fast typist is not a very satisfying communicant, especially since he or she must usually take additional time to think about what to write.

There is a continuum in style of expression exemplified by informal conversation at one end and carefully written, rewritten, and edited composition at the other. Oral and written modes might overlap or alternate. Along the continuum one could place friendly letter writing close to conversation, then real-time interactive typing, small group face-to-face discussion, party-line-type teleconferencing, formal speechmaking, formal business correspondence, Forum-style asynchronous teleconferencing, and finally article and book writing.

Along this continuum style will usually (but not always) change from informal to formal, rough to smooth, and wordy to succinct, and grammar will change from sloppy to correct. Note that the interactive methods are found largely at the beginning of the list, whether they are oral or written. We are not suggesting that this is a vertical continuum, with the worst at the bottom and the best at the top. We would instead consider it a horizontal scale differing along its length as described above, with any point being considered good or bad, effective or ineffective, clear or muddled, boring or dramatic depending on the skills of the particular writer or speaker. The Rudolf Flesch "Fog Index" would, on the average, rise along the continuum I have described.[2]

There are some other features of computer teleconferencing that have been found useful, or that may have a useful potential. It is possible, in Forum, Confer, and various other teleconference systems, to make an anonymous entry. This could be another encouragement to the shy participant, or it could give a person the opportunity to be brutally frank about some sensitive issue that might not otherwise be brought up. This potential for tapping unexpressed feelings has no counterpart in the face-to-face conference. However, during a year of asynchronous teleconferencing in which one of the authors participated, this feature was not seen used for serious purposes.

Another such feature is the private entry, addressed only to some specific participant (or set of participants), but not transmitted to all. Such entries are not recorded in the transcript. This is sometimes described as the "whisper mode," analogous to leaning close to a colleague in a face-to-face meeting and speaking to him or her alone. In the Forum studies at IFTF it was found that some conferences had a lot of private communication within them while others had very little. Private entries averaged about one-third as many as public and were about two-thirds as long. At the Information Sciences Institute teleconference mentioned above, the author was unaware of private entries except the few that he himself made or received.

Since the Forum studies had to be based on real situations, private messages had to remain private and were thus not available for content analysis; all that was recorded was who contacted whom and for how long. Reports from users indicated, however, that private messages tended, as one might expect, to be more informal and chatty, and to contain much more humor than public messages. Almost all the private messages in these studies were between people who already knew each other well. The author's private messages at ISI were often calls for help in operating the system.

One of the most serious problems in computer conferencing is reliability. Because of the complexity of computers, breakdowns are far more common in computer conferencing than in any other medium, including video. If a real-time conference is taking place and the system "crashes" (as computer people say), all participants are suddenly disconnected from each other and the conference stops. Because of this danger, serious teleconferencing in the synchronous mode is seldom attempted, at least with a large group. If the conference is asynchronous, on the other hand, only those participants who happen to be reading or writing at the time are affected. Most users can recall, however, instances of great frustration when, after typing a long and very good entry, the system crashed and lost everything before it had actually been entered into the conference file. These problems are undoubtedly developmental, however; computers are constantly becoming more reliable and soon such things will be of only historical interest.

We have explained why we believe asynchronous computer intercommunication is intermediate in nature between the telephone and written correspondence. Computer terminals (and minicomputers) of one kind or another are rapidly becoming more common in offices and other workplaces, and, as their costs go down, will soon become indispensable in the home as well. It may be purposes other than communication that will motivate this growth, but whenever two people who wish to communicate find that they can do so through their computers, the convenience and inexpensiveness of this medium will assure it an important place in their lives. It is becoming more and more common for secretaries to type their letters and reports not only on paper but also on magnetic tape or directly into the computer in digital form. It is only a matter of time before most letters will be transmitted electronically. When this time comes the postal systems will be relieved of carrying most first-class mail, and will have responsibility only for the delivery of printed materials, books, packages, and advertising material.

notes

1. Alphonse Chapanis (1975) reported on experiments in which pairs of subjects were required to solve problems, communicating the necessary information via various media such as video, audio, telewriting, and teletypewriting. Among other findings was that those media that had a voice link allowed problems to be solved in less than half the time, but required the exchange of around ten times as many words to do it.

2. The Fog Index is a formula to express readability, based on words per sentence, syllables per word, and incidence of personal words or sentences.

5 DATA RESPONSE

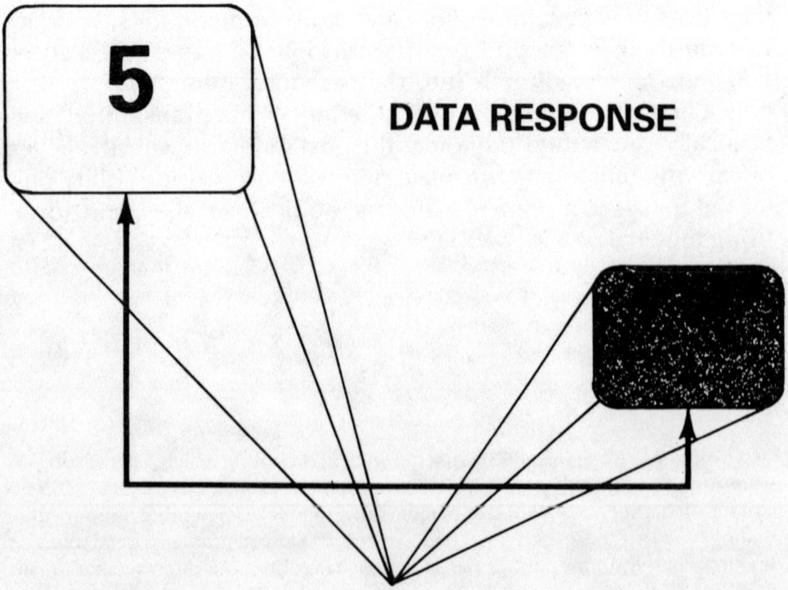

It has long been a dream that we could someday, somehow, invest a one-way mass communication system, such as radio or television, with a two-way interactive capability, so that learners could ask questions and audiences could talk back. It has been done in the case of instructional TV systems by using a different medium for response, such as the telephone or FM radio. This can be practical for systems in which the number of participants is small (even up to 200 in the case of the University of Wisconsin telephone system), but not for the thousands or tens of thousands of receivers in a mass broadcasting medium.

However, a growing number of homes in Europe and North America are receiving their television and radio via wire. A method is now in use whereby the same cable that delivers broadcasting to the home can carry simple data signals in the reverse direction. This very simple data transmission medium, capable of far less information transmission than the lowest-quality voice line, can invest a one-way mass medium with interactive capability. It is the only teleresponse method we have that is practical in mass telecommunication.

The outstanding example of a cable TV system using data response is the QUBE system (Warner Cable Corporation) in Columbus, Ohio. (The success of this system over the last two years has led the company to install similar systems in several other cities.)

The salient characteristics of data response systems are (1) they can elicit and receive response from practically any number of terminals at once, and (2) this information, being in digital form, can be manipulated, processed, and stored by a simple computer. In the case of QUBE, if all 30,000 subscribers should be tuned in at once, and to the same channel, 30,000 responses could be received at once, totaled, averaged, and otherwise manipulated and the results displayed to all participants within a few seconds. In such systems, each respondent has a simple data terminal with a very limited set of buttons, usually in the form of a small box attached to the TV set by a cable of convenient length.

The QUBE terminal had only five buttons; an experimental terminal developed by the Rand Corporation for an interactive instructional TV experiment using a CATV system in Spartanburg, North Carolina, had eight buttons that operated in two modes. One of these modes, in which the buttons were numbered 1 through 8, was used for selected multiple-choice answers to criterion questions posed by the teacher during information presentation. The principle was the same as in classroom response systems such as Edex. The computer processed responses so the teacher had an immediate display of such information as what percentage of the class chose each answer and, if important, just who chose what or who did not select an answer at all.

In the second mode, called the Work mode, the same eight buttons were identified by letter (A through H). The teacher left the system in this mode during presentation so that students could provide constant simultaneous feedback. In the Work mode the A button signified "I understand," and the B button "I don't understand." The teacher could ask from time to time for response using these two buttons. It was like a show of hands in a classroom. If the student pressed the A button without being asked to do so, it would mean "OK, OK, I get it, let's get on with it."

The C button in Work mode had the meaning "Slow down"; the D button, "Give an example"; the E button, "Ask a question, so I can see if I understand." The F button would be pressed if the student thought that the visual the teacher was using was unclear, and the G button was a request to the teacher to call the student on the telephone, when convenient, to explain something that had not been made clear. The eighth or H button was used for logging in and logging out. When a class began, for example, the teacher would ask all students present to press their 8 buttons and, by the resultant roll display, would know just who was present and when latecomers arrived.

Other informal modes were possible at any time; the teacher could assign any meaning to the first two or three buttons, such as "true" and "false," or "easy" and "hard" for response to questions such as "How did you find our last instructional module?"

At the end of the class the teacher had the computer print out a data sheet showing which students made which mistakes on each question asked, and a class summary showing for each student the number correct, the number incorrect, and the number of times he or she did not answer. For the student the computer displayed the answer to each formal question, and the data concerning his or her own responses. There was also the potential for feeding students the group averages so that they could see how they stood among their classmates.

The computer could also print out a summary of responses in the Work mode. An example of such data is given in the Rand report on the Spartanburg project (Lucas et al., 1979: 25). The B and C buttons ("I don't understand" and "Slow down") were very heavily used, A, D, and E ("I understand," "Give an example," and "Ask a question") very little, and F and G ("Visual is unclear" and "Call me on the phone.") hardly at all. This was, of course, only a single example — one session of one course — and may not have been representative.

In the QUBE system the five buttons are used in much the same way, except that the purposes are primarily entertainment rather than instruction. A weekly amateur program, for example, presents performers of various kinds for the acceptance of viewers. During this program the first button means "Yes, let him continue, I like him," while the second means "He's terrible, take him off." It is in the tradition of a nineteenth-century theatrical entertainment called "amateur night," which took place in American small towns. Local talent performed, and if the audience booed loudly and stamped their feet, a gigantic hook came out from offstage and pulled the performer into the wings. Sometimes, instead of the hook, the stage manager appeared brandishing a pistol and "shot" the offending performer. The performers knew the risk they ran, and all was taken in good humor. On the QUBE program, when more than 50 percent of the buttons are pressed for a certain length of time, a horn blows, or a gong sounds, and the performer disappears.

Another regular program features the demonstration and auction of used merchandise. Bids are made by touching button 1. The auctioneer knows when bids are made and how many people made them. He or she raises the price until the number of bidders drops to one, and concludes the sale. The computer feeds out the buyer's name, the seller contacts him or her by phone, and the actual transaction is made in person.

Probably the most practical use of these response buttons is in opinion polling. Publishers have determined the most popular or attractive covers for their magazines among several alternatives. Viewers can be asked to vote on their views before and after a discussion on a controversial issue. For example, in July of 1979, President Carter made an important speech in which he described the energy crisis and made a bid for public support. Immediately following the speech the National Broadcasting Company tied into the QUBE system for an opinion poll of viewers' reactions. Four questions were asked. At the start, the QUBE computer reported that 6500 respondents pressed their buttons; by the third question the total had risen to 7600. This number was more than enough to represent the entire QUBE subscribership; NBC was quick to point out, however, that it could not be representative of the nation as a whole, but could only suggest how communities similar to the predominantly middle-class QUBE service area might respond.

For each question asked, orally and printed on the screen, there was a menu of three possible answers printed below it. The first question asked if people felt more optimistic or less optimistic now about the future. After a few seconds for people to press their buttons, the percentage of respondents making each response choice appeared on the screen: 61 percent said they were more optimistic, 18 percent said less, and 21 percent said they were still confused. To the question, Is Carter being tough enough with his actions? 40 percent said no, 39 percent said yes, 20 percent were not sure, and so forth. Citing "professional pollsters' " opinions that the Columbus sample was indeed representative of the country as a whole, QUBE has continued to make public the results of similar polls on other national issues.

The new QUBE III console[1] introduced in 1980 provided the capability for "teleshopping," that is, purchasing groceries and other deliverable items without leaving the home. Simple banking transactions, such as transfer of funds, and the buying and selling of stocks and bonds was also possible.

QUBE and other systems have been providing monitoring services for the home since 1979. These consist of automatic sensing equipment to inform a central computer of fire or forced entry, plus various push buttons to summon paramedics or police. When this service was initiated in Columbus, Ohio, some 700 QUBE subscribers immediately signed up for it. A major incentive was that the homeowner was given a substantial reduction on home insurance. While this discount amounted to only 5 or 10 percent at the start, it was expected to go as high as 30 percent in short order.

Although only half the size of the earlier QUBE consoles, the QUBE III console also contained equipment for information retrieval, presumably one of the teletext or videotex systems described in Part II of this book. In addition, it was capable of selecting up to 110 television channels in anticipation of future expansion of service.

A cable TV system in St-Hubert, a suburb of Montreal, created considerable interest in mid-1980 with the innovative services it was offering in order to attract subscribers in an area where there was good reception of four American and five Canadian TV stations (see Roberts et al., monthly). It was the first cable TV system to actually offer a channel of video games such as chess, checkers, backgammon, Master Mind, and so on, which the subscriber could play against the cable system computer. This will lead inevitably to the playing of interactive games, not only with the computer, or with other players located in the same room, but also with players located anywhere within the same cable system.

Computer users, particularly those with access to the Plato system, have devised many highly sophisticated interactive games. At almost any time of the day or night, opponents, unknown to each other and located anywhere within the continental United States where there is access to the ARPA computer network, can be found competing in these games. Graduate students in computer sciences and computer buffs in general are largely responsible for the development of these outstanding games, on a hobby basis. Graphic depiction of a simple kind is heavily used, as well as alphameric characters. It is only a matter of time before such gaming will be commonplace in the home.

Of course, interactive gaming is not necessarily interactive communication; what is exchanged in most gaming is responsive action rather than messages.

In order to make possible the kinds of interactions described above, cable TV systems must install special two-way amplifiers in order to carry simple data pulses "upstream" from subscribers' homes to the central computer at the cable system head end. This is relatively inexpensive if done at the time the cable is installed, but prohibitively expensive if done later by replacing the original amplifiers (which are required every few blocks throughout a typical cable system).

Recognizing the potential of data response in increasing the usefulness of cable TV to the public, the Federal Communications Commission (FCC) required that all cable systems larger than 3500 subscribers have the data response capability. The requirement was later converted from "actual capability" to "potential capability." Manufacturers offered amplifiers in which space was allowed for later addition of the return channel equipment, and this the FCC accepted, agreeing that it was not necessary that the equipment actually be in place until needed. Thus all cable TV systems in the United States that have been built, or rebuilt, in the last few years have the potential for adding data response. Meanwhile, a constantly growing number of CATV systems, particularly those in urban areas, are adding data response in order to offer more services and attract additional subscribers.

The reader may wonder, quite legitimately, whether data response in combination with a one-way medium provides true interaction in which each party responds in messages to the other. In the combined one-way TV and data response systems just described, thousands of respondents are aggregated and treated as one. An entire audience may respond to a TV personality, and he or she may respond immediately to the aggregate response they have made. It is nonindividualized interaction, a kind of *mass* interaction if you like, the only case where a mass medium has been used in a real-time interactive manner.

Part II of this book, "Quasi-Interactive Media Systems," describes a variety of data transmission systems that are responsive without being fully interactive. Teletext, in which the user has a small data terminal to select alphameric information from a menu of pages, is an example. All of the systems described in Part II, however, are individualized systems — one respondent (or small group) at a time, interacting in a limited way with a computer.

This chapter has described a number of representative interactive systems using various combinations of audio and visual media. Part III, the last major section of the book, will build on this, going into analytical detail on the problems and issues that face the designer of interactive systems. It will provide a critical assessment of the most common solutions to the various design problems and the theories behind them. But first, a discussion of slow scan TV, and then Part II will carry the description of systems a little further afield, going into a new class of alphameric information systems that are at present only quasi-interactive, but may in the future be combined with fully interactive media or develop fully interactive properties themselves.

note

1. The QUBE console is a hand-held control center with various push buttons for selecting channels, feeding back response, and the like.

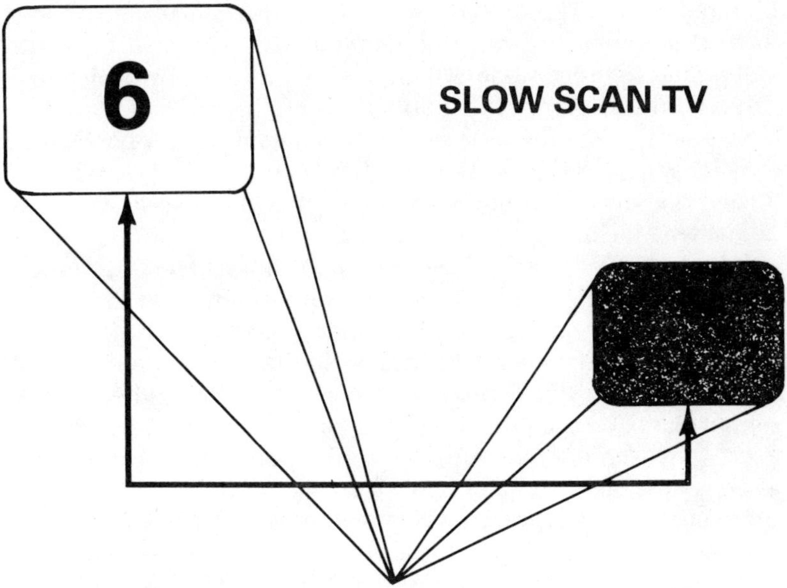

6

SLOW SCAN TV

One of the authors recently visited the Center for Interactive Programs at the University of Wisconsin. In one of the offices a young woman was preparing for a routine video conference. She was talking and listening over a speakerphone on the desk top and checking a series of graphic materials on a TV screen transmitted to her from the other end of the line. The distant participant wanted to check out his materials for legibility before he incorporated them into a presentation. When asked where this participant was located, the woman responded that he was in London, England.

The author had been watching slow scan TV. Quite obviously, a state university would not have been able to afford an international video transmission, let alone a leisurely preconference hook up simply to check out visuals. But this was still-picture video, not full motion, and still pictures can be sent electronically via ordinary voice-quality telephone lines. It was costing no more than an international telephone call. Anywhere that the worldwide networks of telephone wires, microwave beams, and satellite channels can carry the human voice, slow scan TV can carry still

---

pictures. For this reason, slow scan TV has also been called phone-line or narrow-band video and, sometimes, because it transmits still pictures, "freeze-frame" video. Furthermore, any device that can record and play back audio signals can record and play back slow scan TV. It can also be sent over standard narrow-bandwidth radio channels. It is estimated that between 3,000 and 10,000 radio amateurs in over 100 countries now communicate visually as well as audially via inexpensive slow scan TV.

Slow scan TV is to still pictures what television is to motion pictures. Probably the best known example of slow scan TV today is the "Newstime" channel, which appears as one of the cable TV channels in nearly one million homes in the United States. For 24 hours each day, still pictures in color follow each other on the screen at an average rate of one every ten seconds. (News rotates every other ten minutes with other information such as financial, agribusiness, and women's news.)

"Newstime" is distributed via satellite, and while the channel it uses is not as narrow as the usual voice-grade phone line, it is still only about 1/500 of the bandwidth of a standard TV channel. This means that it can be fitted into spare, unused spectrum space beside an existing channel, hitchhiking, as it were, on the transponder that carries the Atlanta super station that is carried by so many cable systems throughout the country. The cost for transmission of this narrow channel is so low that the charge to cable systems for the right to receive and use the service is as little as five cents per month per cable subscriber. TV network news programs, with audiences numbering many millions, charge their advertisers several times this per viewer *per program,* not per month.

TV viewers are also familiar with slow scan TV in the magnificent still images that NASA technicians succeeded in transmitting from the Voyager I and II missions as these spacecrafts passed one planet after another in our solar system.

Slow scan TV, in the "Newstime" example mentioned above, competes with broadcast TV programming as one of many available cable TV channels in the home. There are many uses of slow scan TV, however, that do not attempt this kind of comparison. Medical diagnosis, for instance, is regularly performed via two-way slow scan TV, bringing professional health care to otherwise totally isolated communities such as outposts in the Canadian far north.

There are many applications in which slow scan TV provides an audiovisual communication system between industrial plants or branch offices in various cities so that information in the form of verbal reports or management directives may be quickly interchanged and discussed in joint meetings. A prime example of this is the telecommunications link between Hughes Aircraft Company plants in Canoga Park, California, and Tucson, Arizona. This system incorporates the Darome convener with full duplex audio, NEC (Nippon Electric Co.) black-and-white slow scan TV, the Electronic Blackboard, and high-speed facsimile (20 seconds per page).

The facts that slow scan TV has been slow to develop and its history has been marred by several instances of premature equipment development and marketing failure do not seem consistent with its apparent practicality. Perhaps this is because it has grown in the shadow of full-motion, broadband television. Because of TV there could never be any sort of industrywide agreement that would permit slow scan to take off, so to speak, with manufacturing standards, program production, performing stars, and widespread marketing of the means of reception.

Slide shows were very popular in the 1890s, but disappeared when they could not compete in entertainment value with nickelodeons and Bijou theaters. Had slide projection appeared *after* motion pictures rather than before, it might have faced the same problems as slow scan TV. The nearest thing to slow scan entertainment was the *Women's* program, distributed for more than a year via satellite and cable TV systems. It was a kind of magazine program containing entertainment and informational items of interest to women, entirely in still pictures.

Yet slide shows have certainly not disappeared from the scene. Multi-image presentation has brought the still image back into the theatrical field. In fact, the slide and filmstrip media, especially when integrated with audio, are today the fastest growing segment of the entire nontheatrical audiovisual field, as demonstrated in the Thomas Hope reports of recent years. There clearly is a place in the mix of modern media for still pictures.

Since slow scan TV is a kind of television, it invites comparison with full-motion TV and, on the basis of effectiveness only, appears inferior. If it were compared to the audio medium, as it often is now by people planning to build teleconferencing systems, it is seen in a very different light. With teleconferencing in mind, slow scan TV has sometimes been compared to facsimile. The relationship is distant, but the two media often become alternatives for the augmentation of audio systems. Both transmit visual materials over very narrow and inexpensive channels, but there the similarity ends.

Facsimile produces a "hard copy" on a piece of paper. It can reproduce far more detail than slow scan TV; a page of typewriting can be reproduced, for example, or highly detailed charts, tables of figures, and the like. But facsimile is slower and can transmit only flat copy; there is no standard facsimile *camera* that can originate pictures of people, for example, or objects larger than might be placed on the pickup stage of a photocopy machine.

By no stretch of the imagination can you consider facsimile a live medium operating in real time, as live television and radio do. It is only a means of transmitting previously recorded pages. Slow scan TV, on the other hand, *can* transmit a still picture the moment it is "taken," and although it may require some seconds before the transmitted picture is complete, it still constitutes a look at something happening in a distant place at the time of viewing.

Most important, slow scan TV can transmit essentially everything that TV can transmit, without, of course, the animation or motion component. In most slow scan systems closed-circuit TV cameras are used at the sending ends to originate the images, and standard TV screens are used at the receiving ends to display them. In addition to graphic materials, the camera can view subjects from the microscopic to the macroscopic, flat or three dimensional, in monochrome or in color if a color system is used (color facsimile will soon be on the market).

Still Images as a Means of Visual Communication

Are still pictures less useful than motion pictures, or are there purposes for which stills may be as good or better than pictures in motion? There is some rather convincing evidence of the equal usefulness of stills from the field of instruction. Researchers have asked: Is motion really *used* for instruction when it is available? At least three times in the past, at about ten-year intervals, someone has undertaken to study the leading educational films of the time to determine just what proportion of such films contained motion that was relevant to their apparent learning objectives. This eliminated background movements of people, traffic, and the like, and camera movements that did not appear to have any direct or indirect instructional purpose. In each of these studies, a core of less than 40 percent in any film was found to contain relevant motion; 60 percent of what such films conveyed could have been (and often was) expressed in still pictures. The next logical question to ask was: How important is motion to instruction — does it actually improve learning?

Several experiments have been made involving, in all, literally hundreds of college students, student nurses, and military trainees, in which the amount of learning achieved from films was compared with that resulting from the same objectives taught by slides or filmstrips. In all cases there was no significant difference in learning.

An equally important use for slow scan TV is teleconferencing. Two or more persons, distant from each other, engage in conversation. The content of this, as in any conference, lies in what the people say, although users of audio-only teleconference systems often feel limited when graphic materials or other visual aids are needed. Everyone seems to feel that relating to a person at the other end of the line is easier, and hence expressing oneself verbally is easier, if one can see the person one is talking to. Naturally a motion picture of this person is more realistic than a still image, even though all he or she may do is blink occasionally and display very subtle changes of expression while listening.

In designing a teleconference system, a major objective is to create the effect of "social presence," that is, the feeling in the user that he or she is talking with someone who is present in the same immediate space. Few designers have had much faith in the powers of the still picture to generate or enhance the feeling of social presence. In most instances where slow scan TV systems have been built for teleconferencing purposes, slow scan has been used only secondarily to convey images of people; its purpose has been mainly to convey images of graphic materials, drawings, print, photos, or slides — in short, visual information. Social interaction has been left almost entirely to the audio component of the medium. (Some audio systems, such as the Remote Meeting Table system of Britain, can be quite successful in establishing the feeling of social presence, through the use of some sophisticated augmentation short of transmitting pictures.)

During the mid-1970s, RCA Global Communications marketed for a brief period a slow scan TV system that they called Video Voice. The user terminal looked much like a Picturephone, consisting of a desk-top unit with a seven-inch monitor and a camera just above it directed normally at the user but with a pull-out mirror for viewing pictures or documents. One of the authors used this system to converse with someone in another city whom he had not met and found the experience surprising. Although a new picture of the man did not appear more often than about every thirty seconds, each showed him in a somewhat different attitude, with a different expression, and added measurably to the impression received from how he talked and what he said. Full and realistic motion, or full and realistic color it wasn't. But it was much better than audio alone in giving a feeling about the man, and this had not really been expected.

Slow Scan TV Systems and Their Users

A few words explaining how slow scan systems work and how they appear to their users will be useful here. The visual component of most slow scan systems today starts and ends as regular closed-circuit video. The audio component, as it is in regular television, is a separate but parallel medium essentially no different from other audio systems, which of course vary widely in quality and sometimes employ dual systems for stereophonic effect. In the visual channel, on the other hand, a mixture of media components is used. A standard closed-circuit TV camera picks up the image to be transmitted and a standard TV receiver or monitor displays it at the receiving end. Sometimes, as we shall see, more than one TV screen may be employed for display. The slow scan feature, then, concerns only the intermediate transmission system between the camera at one end and the TV screen(s) at the other.

Regular television transmission, by contrast, is "fast scan." An image created optically on the face of a camera tube is dissected line by line to create one long, continuous electrical current, modulated according to the light or dark areas that the scanning electron beam encounters. Thus a two-dimensional image is transformed into a one-dimensional current (called the signal). This is what "scan" means. The beam sweeps back and forth until it has covered the entire picture area.

In order for one manufacturer's receiver or monitor to display the signal generated by another manufacturer's camera, scanning standards were established, of which there are several in various parts of the world. Most standard systems scan the picture with horizontal lines, 525 in the United States and Japan, 625 in most European countries. Some nonstandard systems scan with vertical lines; a few use a spiral scan, covering the entire picture, in effect, with one line.

The U.S. television standard specifies a whole new picture (called a frame) thirty times per second. On the receiver screen these still pictures, like film frames, follow each other so rapidly that they blend into a continuously moving image. The inner surface of the picture tube is coated with a phosphor that glows as the scanning beam moves across it. There are phosphors that glow for several seconds, such as those used in radar scopes. In the TV picture tube, however, the glow must disappear in one-thirtieth of a second or any rapidly moving object, such as a hand waving, will leave a trail behind it. We call this one-thirtieth of a second the "update" or "refresh" time. If another scanning does not immediately follow, the screen will go black.

Radar scopes are updated every few seconds by a beam that scans in and out from the center, sweeping around in a circle to cover the entire circular area of the scope in the update time. Slow scan systems have a much longer update time — from a few seconds to more than a minute. There are no inexpensive TV monitors on the market that, once scanned, retain an image for a minute or longer (although early slow scan equipment did utilize such equipment). Inexpensive closed-circuit TV monitors, however, are available off the shelf in great variety. If slow scan pictures arrive at a terminal at the rate of one every thirty seconds, but a closed-circuit TV monitor that requires updating every one-thirtieth of a second is used, an interface problem arises. The problem is solved by incorporating into the receiving equipment a memory system that records at the slow scan rate as the picture comes in, but plays back at the TV refresh rate for continuous display.

With few exceptions slow scan TV is not at present used as a "program" medium as regular television is — that is, people don't plan or write slow scan TV "programs." The closest we have come to this is in the pick up and transmission of lectures and other presentations. The use of slow scan as an aid to audio teleconferencing is perhaps another example.

The simplest and least expensive slow scan systems use a simple telephone circuit for transmission, and display an image as it is being received. A meeting will begin in the audio medium only; when a picture is to be sent, both participants stop talking, switch to video transmission, and watch the transmitted picture take form. Starting with a blank screen, the picture builds up, line by line, gradually filling up the empty screen. A low-definition picture can be transmitted in eight or nine seconds; a very high-quality picture can take over four minutes. The usual update time of thirty seconds provides a fair-quality image, comparable to the quality with which average home viewers are familiar on their not perfectly maintained or adjusted home receivers. There are some obvious drawbacks to the slow build-up effect for applications such as instruction and teleconferencing.

When telemedia are used in instruction, it is usually either to bring instructional presentations (lectures or demonstrations) to the distant learner or to bring teacher and learners together in an interactive mode for discussion, student response, evaluation, or guided performance. Lectures and demonstrations are essentially one-way and require a medium capable of continuous and unified presentation. Cessation of the verbal lecture and a wait for the transmission of visual materials can be disruptive to this program flow.

In regard to the interactive methods of instruction, the media characteristics are generally similar to those required for teleconferencing. If only two persons are involved in interactive exchange, the situation is sufficiently informal that waiting for the picture to be transmitted is tolerable. When there is a group at either end of the line, and particularly when two or more groups are involved, at different places, uninterrupted contact is fairly mandatory to keep the meeting in order. Audio need not be interrupted if two telephone circuits are used, but this of course doubles the cost of transmission. One circuit is used in the normal way for continuous interactive audio. The other may be used, as needed, for picture transmission in one direction or the other (but not in both directions at the same time).

When the application calls for the continuous transmission of still pictures, as in the case of "Newstime," or a teleconference in which pictures of the participants are transmitted, most slow scan systems can maintain a continuous picture by building up the new picture while simultaneously erasing the old one. This is done line by line; the line of demarcation between the two pictures moves across the screen as a wipe. The "Newstime" wipe is a vertical line moving horizontally from left to right, because in that instance the scanning lines are vertical. The NEC system transmits and displays a color picture in a similar fashion, but with a vertically moving wipe because NEC scans the picture with horizontal lines. If a series of pictures of the same subject are being transmitted, as, for example, pictures of participants in a teleconference, only those areas that have changed from one picture to the next betray the presence of the wipe line; it is not visible as it moves across background areas or objects that have not moved during the update time.

The more sophisticated slow scan systems eliminate the wipe in favor of an instantaneous cut. To do this a second memory is incorporated into the receiver, or the first is extended until it is capable of recording a second full frame. After the first picture has been transmitted and locally recorded (which can have taken place before the beginning of a presentation) it is scanned thirty times per second to refresh the TV monitor image, while the next frame is slowly being transmitted and recorded on the second frame store. A remote control is provided so that after this second transmission is complete, and the sender is ready to use it, he or she may send a pulse to switch it onto the receiver display. The first frame store is then free and can be erased and used to record the third transmitted frame, and so on.

The sender must obviously delay talking about the new image until it is complete and displayed. Colorado Video systems provide the sender with a "cursor" on the monitor. This is a white line that travels across the picture (a vertical line in this case, because vertical scan lines are used), telling the sender the degree of completion of the transmission as it moves from the left-hand side of the screen to the right.

One of the problems encountered in the design of slow scan systems is keeping a person still while a long exposure, so to speak, is transmitted. If a person moves during transmission, part of him or her may be shown in one position while another part, scanned later, may be in a different position entirely, causing a strange distortion of form. Unless people can be relied upon to hold still, which is impractical, such systems are usable only for inanimate materials such as photos, drawings, objects, and the like.

The solution to this problem has been to incorporate some means of "freezing" the picture before it is scanned and transmitted. This is now done with a solid-state memory that, on the sender's "take" signal, records one television frame. From there the slow scan transmitter system sends the frame one line at a time at whatever rate the transmission channel bandwidth will allow.

Instead of sending thirty pictures per second, as we have said, slow scan sends one picture every thirty seconds (for comparable picture quality), which amounts to sending only 1 frame out of every 900. Roughly, we could say that that would require only one-nine-hundredth the bandwidth, although the difference is greater than that in most cases. Since slow scan starts and ends with standard TV, the process is often referred to as "compression"; the sending equipment, with its freeze-frame store and scanning system, is called a "video compressor"; and the receiving system, which does the reverse of this, is called the "video expander."

It was mentioned earlier in this article that many radio amateurs are using slow scan TV. Except for certain channels, a ham operator does not need to make any change in his or her existing transmit and receive equipment. A visual transceiver (marketed by Robot Research), called a "scan converter," needs to be added; currently these cost around $800. Most operators adapt small TV receivers as

TABLE 2 Relation of Picture Quality to Frame Frequency

Pixels[a]	*Definition in Legible Characters/line*[b]	*Frame Frequency*[c] *(in seconds)*
4,096 (64 × 64)	10	2
16,384 (128 × 128)	20	8
65,536 (256 × 256)	40	35
262,144 (512 × 512)	70	140

a. Pixels are picture elements. Because the slow scan TV aspect ratio is the same as that of TV—four wide to three high—each of these pixels is somewhat elongated horizontally.
b. Picture quality in terms of the maximum number of characters in a legible line of print (within the safe or usable area of the picture — 90 percent of full picture width).
c. Shortest possible time between full frame displays (same as update time).

their visual monitors, at little if any additional cost. Successive pictures wipe on from the top down with a horizontal line. Since a ham has only one transceiver, he or she can either talk or send a picture, or either listen or receive one, but cannot do two of these things at once.

A sequence of pictures is transmitted at a fixed and relatively high frame frequency of eight seconds per picture. Because the picture must be transmitted on a voice-grade channel only one KHz wide, the quality of the picture is fairly coarse (see Table 2). While hams with such equipment are estimated to number between 3,000 and 10,000, this constitutes only 2 or 3 percent of all radio hams. In order to concentrate their efforts and increase the probability of making visual contact, these specialists generally "meet" on Saturday afternoons, eastern U.S. time, to exchange pictures with respondents in all parts of the world.

Trade-Offs: Picture Quality, Frame Frequency, and Bandwidth

Different applications for slow scan have different requirements. There are some cases in which detailed engineering or architectural drawings must be discussed, and in these cases slow scan TV alone is simply not suitable. It must be augmented by the use of other higher-definition visual media, such as facsimile. The legible transmission of the full width of a typewritten page (60 to 70 pica-size characters and character spaces) cannot be satisfactorily achieved with commercially available slow scan equipment — or regular TV, either, for that matter — let alone wide-page printed materials with a smaller type size. Two-column typing, however, is no problem for the higher-definition options of slow scan (see Figure 18). Obviously, a system that is to be used mainly for the transmission of detailed graphic materials should emphasize picture quality.

On the other hand, a system that is to be used for teleconferencing may have limited definition and still be completely adequate for pictures of people. Close-ups, as a matter of fact, are often better if the images are somewhat less clear, as evidenced by the fact that Hollywood film makers used to use diffusion filters over camera lenses on close-ups to deliberately reduce the definition of skin texture or blemishes and flatter the stars. Of course, in this example we are talking about a much higher range of definition. However, when the choice is between high frame frequency, such as eight seconds, and better definition, and the subject is people's faces, definition is generally less important than frame frequency.

Ideally, a slow scan TV system should have a short update time and a high-definition picture, and should require a very narrow (and inexpensive) transmission channel bandwidth. However, a designer can maximize one of these factors only at the expense of one or both of the others. A one-KHz channel can transmit 2000 picture elements per second. Starting with an average picture definition, update time, and channel bandwidth, the trade-offs are as follows:

(1) If the designer would have a higher-quality picture, a longer update time or a wider bandwidth (or both) must be accepted.

(2) If the designer would achieve a shorter update time, a poorer-quality picture or a wider bandwidth (or both) must be accepted.

(3) If the designer would transmit over a narrower bandwidth, a poorer-quality picture or a longer update time (or both) must be accepted.

The designers of the "Newstime" system (Colorado Video), wanted the news programs to move along at a fairly brisk pace. After initial research, they decided that a picture update time of eight seconds would be generally acceptable. At the same time, they were not content to accept a picture quality that compared unfavorably with the definition that viewers were accustomed to in watching broadcast television. They wanted *both* a short update time and a high-quality picture. As a result, they could not use voice-quality channels for transmission and had to go to a ten-KHz channel (a bandwidth generally used for high-fidelity music transmission).

There is still a fourth variable that may be altered in some special cases in order to improve one or more of the other three. This is the aspect ratio (and overall size) of the image. If a smaller area of picture is required, with parts of the receiving screen left blank, there will be fewer picture elements to be transmitted for the same picture quality, and either a narrower bandwidth or a shorter frame frequency can be achieved. Alternatively, the same number of pixels may be concentrated into the smaller area, allowing for a higher-quality image. Robot Research produces a slow scan signature verification system in which no more than a horizontal band of picture, one-fourth of the full screen height, is transmitted, large enough to show a signature, into which the full number of pixels is concentrated.

Origination and Display Techniques

Most slow scan applications have been relatively simple, both in origination of the picture and in its display. The purpose of using slow scan at all is to find an affordable means of transmitting visual materials over long distances, and making a system more sophisticated naturally adds to its cost.

So far this discussion has been predicated on the concept of a single system, camera at one end and monitor at the other, with the implication that if such a medium is to be interactive (like the telephone) there will of course have to be a second such system

20 characters per line

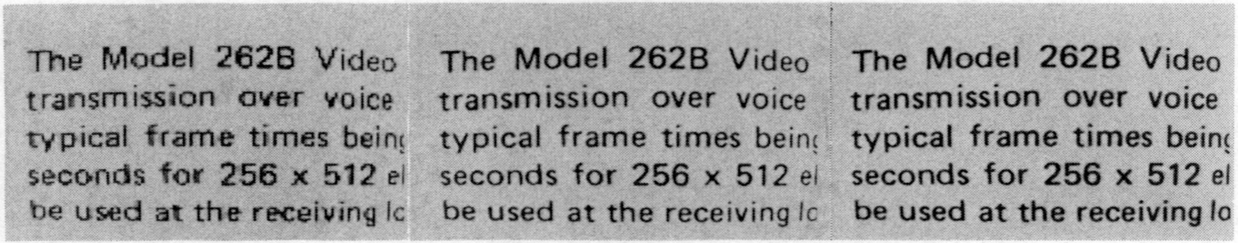

40 characters per line

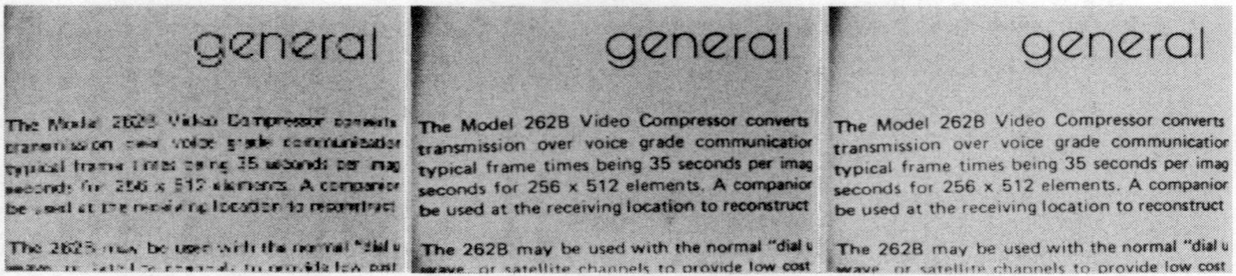

80 characters per line

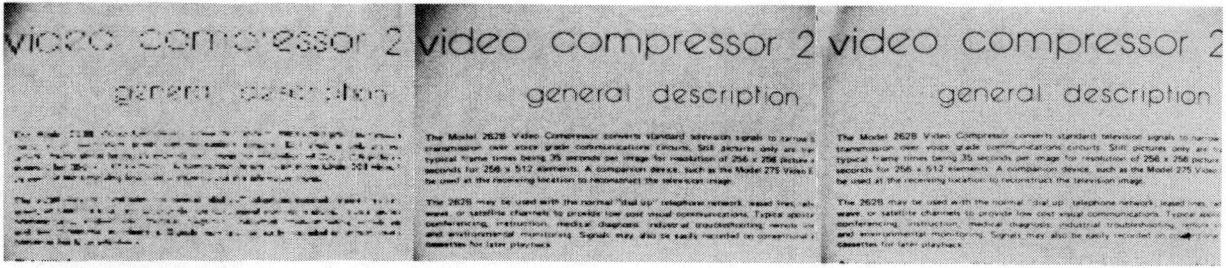

128 × 128 (16,384 pixels) 8 seconds	256 × 256 (65,536 pixels) 35 seconds	512 × 512 (262,144 pixels) 140 seconds

NOTE: Adequate legibility is represented by the upper-left, center, and lower-right images. The three images at the lower left are not legible; the three at upper right have greater definition than necessary. Photos courtesy of Glen Southworth, Colorado Video.

Figure 18: Legibility of three sizes of print when digitally transmitted at three frame frequencies (thus three levels of definition).

working in the reverse direction, a camera associated with the first monitor and a monitor associated with the first camera.

When the system is used to augment a verbal discussion, a picture may be sent now and then, with no attempt to keep a continuous flow of pictures on the screen. In more complete applications, using a separate phone line for picture transmission, pictures may be sent much more often, with the attempt to always have something relevant on the screen. If the wipe transition is via a horizontal line, coming down the screen from the top, it will reveal text line by line in the familiar fashion of titles rolling up the screen. Items are revealed one after another and, with proper pacing, can be read aloud off the screen as they are revealed. This gradual revealing is generally considered better teaching strategy than showing a list all at once, since all attention is generally centered on an item just revealed. Perhaps in the future a speaker will be able to control the sending of the picture so that he or she can send a line or so, then stop sending until discussion of the item is complete, and when ready send another few lines and reveal another item in the list, thus building up the picture at the user's own rate of speaking.

There is another possibility that has considerable instructional value. Once a picture has been transmitted, the sender may superimpose a pointer on the picture and move it remotely, directing the viewer's attention to specific locations as they are discussed. Simple pulses to move the pointer up, down, left, or right are transmitted on the picture channel. NEC provides an "electronic arrow" for this purpose; other systems for pointing have used movable cross hairs or small oblong cursors such as those that word processors employ.

For greater flexibility the standard techniques of television production may also be used, involving two or more cameras and a switching system to select among them. Unlike regular television, the receiving end may be provided with multiple display screens, each displaying a different picture.

In the case of a teleconference, for instance, three cameras may focus on two people each, with a fourth mounted on a graphics stand for close-ups of flat materials and small objects. Slides may be readily incorporated using one of the simpler systems found at cable TV control centers, in which a slide projector is aimed directly into the lens of a special TV camera, or, more simply, the slides are projected onto a screen and the screen image is picked up by an ordinary TV camera. In the case of medical "grand rounds," in which a series of patients are presented and discussed, one camera may be focused on the doctor who is presenting a case along with the patient he or she is describing, a second may be used for close-ups of either doctor or patient, a third on displays of the patient's X-rays, cardiograph record, or vital signs, and so forth. If the cameras are manned, one camera may serve several purposes, far more so than in standard TV production, since only one frame is required for each shot the camera takes. If the system must be automatic, the number of subjects that can be covered will be restricted, and so will the movements of anyone before the camera.

Thus at the origination end, a complex slow scan TV system can look very much like standard television. At the receiving end, however, things can look altogether different. Whereas standard television involves the sequential transmission of a shot from one camera after a shot from another, and their sequential display on a single monitor, a series of slow scan shots may be recorded at the receiving end and displayed on several different monitors. In the example just given, for instance, the patient may be shown on one screen, X-rays on another, vital signs on a third, and so forth.

Advantage is thus taken of the fact that a still picture represents a frozen moment in time — it is timeless, in that sense — whereas motion is time based. Movies and TV move through time, or "run" as we say, which is probably a reference to the film projector, which, like any machine, operates in time. Movies and TV have a finite time dimension; still pictures have no time dimension unless it be merely implied. Of course, reality has all the dimensions, and a conference or medical demonstration does move forward in time and can be represented adequately by any given still picture or set of stills for a limited period only. A close-up of a graphic illustration that is no longer under discussion and irrelevant to what is being said is only a distraction. In such cases the system design may allow the sender or the viewers to replace it with something more recently transmitted. One custom-designed Colorado Video system provided the capability of recording up to 27 pictures at the receiving end; another allowed random access to as many as 500 images.

It is true that pictures in motion are far more realistic than still pictures. It takes more than the factor of motion, however, to make a medium suitable for teleconferencing. The essence of the teleconference situation is that participants are *present* simultaneously — it all happens in real time. If it were not so, they could not respond to each other, back and forth, in the conversational mode. Motion pictures and videotape, being recording media, not telemedia, are impossible to use in interactive communication, however realistic they may be in motion, color, and even three dimensions.

Slow scan TV, however, is a telemedium. It is a means of live transmission. Thus it shares that characteristic of standard television that makes it so useful for teleconferencing: It conveys the sense of immediacy — the feeling that what you are seeing is, *almost* at that very moment, happening in a distant place. We must say "almost" because any slow scan image is of course a few seconds old by the time it is transmitted and displayed.

Over the Horizon

It is conceivable that some future slow scan system might provide even more flexibility by simplifying the origination process. With a multiframe store at the origination end as well as the receiving end, a single camera with camera operator, zoom lens, and flexibility in moving could provide a large variety of still pictures in rapid succession — wide shots, close-ups, different camera angles, and so on — just as multiple cameras do in a closed-circuit TV application. The camera operator would simply go from subject to subject, taking "snapshots" so to speak, of all likely subject matter. A switching operator, watching four monitors, say, like the four monitors at the receiving end, plus a preview monitor or two, could choose the order in which images are transmitted, retrieving appropriate shots from local storage as needed. The center two monitors might display shots of the people in a teleconference, a side monitor graphic materials, and the other side monitor special close-ups. Four switching buttons would select the image to be transmitted after it had been previewed; it would be transmitted along with an identification so that it would automatically appear on the appropriate monitor at the receiving end.

Developing Intermediate Stages on the Still-Motion Continuum

Still pictures and pictures with full and realistic motion are all really still pictures; they differ only in frame frequency. Between the printed picture at one end of the scale (which never changes, and hence has a frame frequency of zero) and the TV image (which updates thirty times per second) there is a continuum. Silent films were standardized on a frame frequency of 16 frames per second. With the advent of sound on film a new standard of 24 frames per second was established. This was simply because, in 1927, 35 mm film running at 16 frames (1 foot) per second, did not move fast enough through the sound head for good sound reproduction. Today better-quality sound can be reproduced from tiny audio cassettes running at a quarter that speed or less, but the 24-frame-per-second standard remains. That high a frame rate is not really necessary for smooth motion. Animators rarely draw 24 separate pictures for a second's worth of action; 12 drawings, 1 for every 2 frames, or less, is usually satisfactory. Fewer than 8 per second may begin to look a little jerky, especially in moments of fast action, but still gives the semblance of lip sync. At 4 frames per second, it looks less like motion than like a series of still pictures. Viewers are quite used to the effects that result all along this continuum. During the NASA Apollo mission, which is now over a decade in the past, color TV broadcasts were fed to the TV networks from the space capsules en route to the moon. Because broadband channels were not available for this, narrow-bandwidth channels had to be used, with consequent compromise in frame frequency. Yet these transmissions were accepted as live television, especially since they were accompanied by what was obviously live sound.

Going back even further in television history, before the era of satellite transmission, there was a period of a few years during which news film was transmitted from Europe via undersea cable, using existing audio channels. Here again the necessary slow frame frequency caused a jerky effect that was acceptable because it provided a means of seeing current film from Europe. Better-quality cable film was possible if it could have been transmitted more slowly, but time for this was not available. Competitors were rushing regular newsfilm across the Atlantic in specially commissioned planes, developing and editing en route, and transmitting from the Montreal airport, with no quality or frame frequency limitations and no particular limitations on film length.

Nonredundant Transmission

An even more practical possibility is that those parts of a picture that do not change from frame to frame (such as background elements, for example) need not be retransmitted in each frame but can be simply retained in the display until they do change. This can cut way down on the bandwidth requirements, depending, of course, on the amount of action transmitted.

Considerable development is under way through new means of digital encoding to reduce the bandwidth requirements of full-motion TV. It would seem that after the necessary circuits have been designed for that purpose, where the need and the payoffs are greater, the same concepts may later be applied to slow scan TV. In this case, however, it will be applicable only for teleconferencing, for example, where a series of images of the same person are to be transmitted, differing only in expression, tilt of the head, or sometimes in gesture. The usual series of graphic materials or other visual aids and illustrations exhibit no redundancy between one picture and the next and generally require a high degree of detail.

These developments will encourage use of the middle range of the still-motion continuum, reducing the bandwidth (and hence the cost of transmission) of pictures that seem to be in motion, and increasing the quality (or shortening the frame frequency) of still pictures transmitted via slow scan TV.

QUASI-INTERACTIVE MEDIA SYSTEMS

II

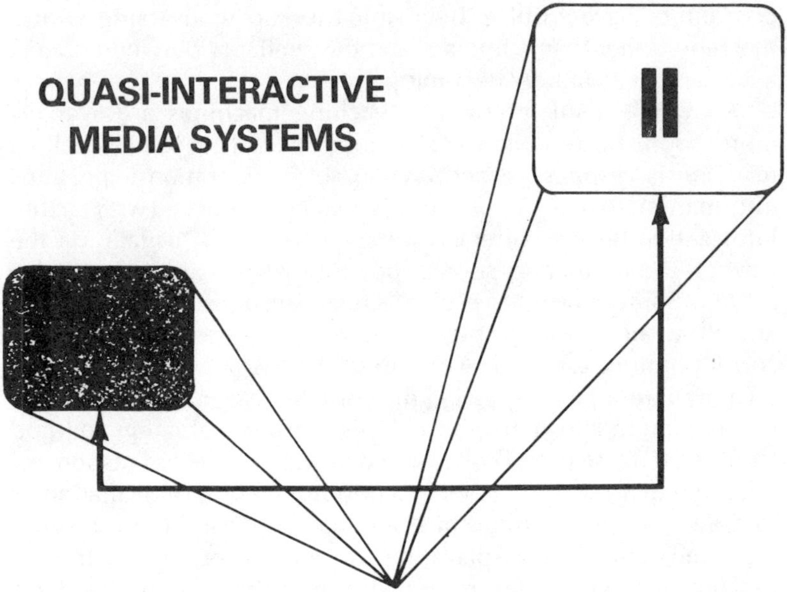

The term "interactive" is frequently applied to systems that are not fully interactive according to the definition accepted for this book. An interactive communication, to review our definition, is one in which each of two (or more) communicants receives and responds to messages originated by the other(s). This means that the system is functionally symmetrical; the roles of the communicants are interchangeable; each may alternately be sender and receiver. A "communicant" may be either a human or a sophisticated machine, such as an elaborately programmed computer.

The average person interacts daily with simple machines that perform transactions of various kinds, such as giving change for a dollar bill and dispensing coffee and other food and drink — sometimes in response to a large variety of orders (black, with cream, with extra cream, and so on). But the roles are not interchangeable — the coffee machine doesn't order anything from the human or even ask what he or she wants. The well-programmed computer, however, does ask questions and responds in a specific way to the answers, and, for instructional purposes, for instance, does what a human tutor might do. Even in its current stage of development, a properly programmed computer may be considered a communicant. Some first-time users of sophisticated computers have trouble believing they are conversing with a machine rather than a human, and the similarity between human and machine is rapidly becoming greater.

Many so-called interactive teaching machines are scarcely more sophisticated than coffee machines. The usual teaching machine is response paced. When set in operation it presents information, then asks a question, refusing to proceed with further information until a correct response is received, usually via the correct choice among several possible responses that are displayed. Some machines evaluate these responses automatically and allow the learner to proceed after he or she has pressed the correct button. Others leave it up to the learner to do his or her own evaluation, merely giving the correct answer after the learner has committed to a response. This learner-evaluation method lends itself to the use of constructed as well as selected responses. A learner may be asked to write a description or an explanation or to draw a simple diagram. In some machines, usable for testing, the same action that displays the correct answer (usually the advancing of a paper tape roll) also moves the response under a transparent cover, where it is still visible but inaccessible for changes.

Such teaching machines are not interactive in the full sense, since they do not respond *specifically* to what the learner does: In the case of most automatically evaluating devices that restart when the correct button is pushed, the next presentation is the same for any learner.

There are fully interactive teaching machines, however, that respond specifically to a certain range of learner responses. These are automatic-evaluating devices, and their responses, all preprogrammed of course, take account of various learner difficulties as revealed in the particular wrong answers that have been chosen. The machine may then either review past presentation (content that the student has shown he or she did not learn) or lead the student into a special remedial sequence that may attempt to explain the same information more slowly or in a different way. These procedures are known, generally, as "branching." Some branching systems take the fast learner into account and may, as a result of a learner's demonstrating prior mastery of what is to come, skip over material that he or she already knows. The most common examples are the modern computer-assisted or computer-managed instructional systems (CAI and CMI). These systems are fully interactive according to our definition, at least in the highly sophisticated examples, although there have been many instructional uses of computers that were no more interactive than the simplest teaching machine. To the charge that the interchange of roles is more illusion than reality, since all of a computer's responses must be programmed into it in advance, it can be said that much of human interaction is no more personal. For example, the responses of government workers to questions from the members of the public they serve are, often by law, limited to what has been programmed for them in their rule books.

But the human has *potential* for transcending his or her programming, on the basis either of reason or of emotion. The machine makes judgments or decisions only on the basis of its program. The computer's brain has only one "hemisphere," analogous to the human left: intellectual, linguistic, mathematical. The human has also a right hemisphere: nonverbal, emotional, geometrical. In a normal human, these functions interact; a person must exert great effort to prevent the feelings of the right brain from interfering with the logical thinking of the left, if indeed it can be done at all. Thus human logic is frequently imperfect, whereas computer logic is rarely so.

Storage And Retrieval Systems

Another kind of system that is often called interactive, but does not fit the full definition, is the system for central storage and remote retrieval on demand. These are essentially selection systems, in some cases assisted by a sophisticated index. The user (who should not be called "communicant" in this case) is primarily interested in receiving prepared programs or specific information, and a typical system allows the user to send simple pulses in the reverse direction to make a selection at the central location. The user may be a teacher selecting specific films or tapes for classroom use — instead of having to schedule them days or weeks in advance and operate a film projector in the classroom, they are played back at the central point and screened on a classroom TV receiver. (When selection of this sort is made by telephone, ordering through a clerk, the telephone is of course an interactive medium, despite its being used for the limited purpose of selection.) Or the user may be an individual at home, accessing pages of information for display on a home TV screen such as might be found in a yearly almanac, except that the data are constantly updated.

The cable TV system in St-Hubert, Quebec, provides an example here as well: It is the first cable TV system to have offered an on-demand program service to the home. Choosing from a catalog of some 5000 titles, the subscriber calls in a request, and (if the program he or she wants is not already scheduled on one of the six channels devoted to the service) the central controller schedules it at once, at a time convenient to the caller. Prime-time periods, when the demand is heaviest, require a "voting" procedure, in which programs that accumulate the most votes are the ones to be shown; others must be scheduled at less convenient times.

None of these examples constitutes full interaction, since (1) the sender-receiver roles are not interchangeable — receiver remains receiver throughout, and the simple pulses returned for selection purposes cannot be considered messages; and (2) the sender is not a communicant, and does not react to an order received by originating and sending a message, but only by a simple action. An intelligence capable of comprehension, evaluation, and construction of a response is not required.

In this part of the book, we will discuss several kinds of quasi-interactive information retrieval systems designed to incorporate the home television receiver as part of the system.

Most homes in the western world are equipped with one or more television screens; in Canada and the United States, practically all homes are so equipped. During the middle and late 1970s, electronic firms in several countries were developing additional ways to use the television receiver, and the broadcasting and communication organizations were devising new services to exploit these possibilities. Several kinds of systems were developed that could provide pages of printed information, plus more or less elaborate graphics (diagrams, graphs, maps, drawings, and so on).

By the end of the decade, three different types of information services were being offered in several places in the world. We will discuss two of these. The first, "teletext," was an auxiliary broadcasting service that was fed to the home as part of a standard television signal, without affecting the TV broadcast. With a selector switch the viewer could choose one or the other, or both at the same time if, for instance, he or she wanted to read the latest news flashes or market reports on the screen without leaving a football game. Another example of teletext was Reuters IDR, a cable television service that utilized a full television channel and hence had a much greater capacity than an auxiliary broadcast service. The second service we will discuss, "videotex," was not broadcast but came into the home via the existing home telephone line, using the television only as display screen.[1] There was only a practical limit to the information capacity of this kind of system; there was no theoretical limit.

note

1. The British use the term "viewdata" to cover information transmitted to the home by telephone wire, leaving "videotex" as a generic term to cover both teletext (as described here) and viewdata (which we are calling videotex).

7 TELETEXT AND REUTERS IDR

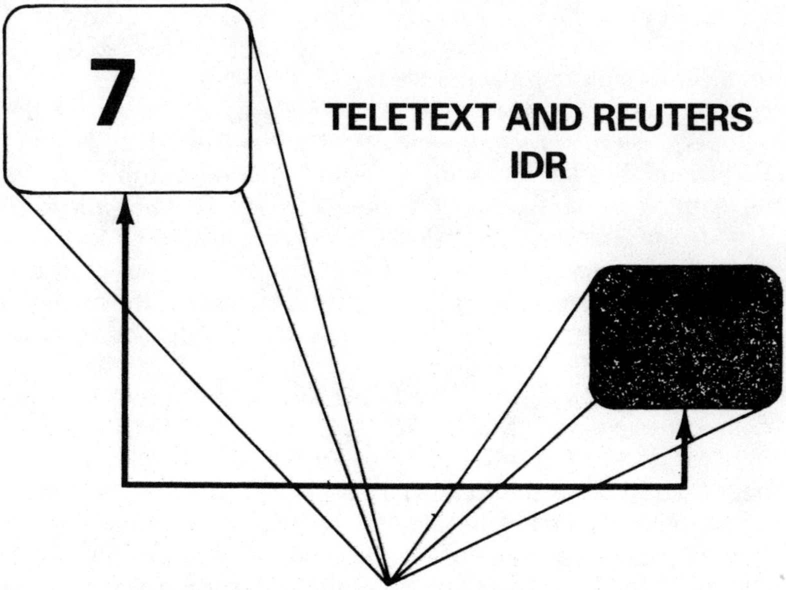

The auxiliary broadcasting services that have been known generically as "teletext" systems were developed originally in Britain and France. It had long been realized that there were a few scanning lines in the TV signal that were not being used to carry picture and sound, lying between the end of one scanning field and the beginning of the next, when the scanning beam retransverses the picture area to begin again scanning the new field. There are at least two such lines in the vertical retrace interval between fields that engineers have always felt might someday prove useful. RCA once designed a facsimile service to be transmitted on these lines, but never brought it to market. In 1974 both the BBC (British Broadcasting Corporation) and the Independent Broadcasting Authority (IBA), Britain's commercial network, developed information dissemination services using these lines.

Although somewhat different, and hence incompatible, these two systems operated on essentially the same principle. The BBC system was called CEEFAX (a cute way of spelling "see facts") while the IBA system is called ORACLE. IBA certainly deserves some sort of prize for the longest reach in deriving a relevant term from the letters of a noble-sounding acronym. ORACLE stands for Optional Reception of Announcements by Coded Line Electronics. The need for standardization was of course realized by both organizations — as the need for any broadcast standards is recognized, so that all receivers can pick up all broadcasts — and by 1979 they had found a common ground. Meanwhile, the French government had developed a similar system, which they entitled ANTIOPE. By the middle of 1979 several additional countries were known to be developing teletext systems of their own, or adapting those of Britain and France to their local requirements. These included Australia, Germany, and the United States. Finland, Sweden, and Japan had also developed information systems to use the home TV screen, but these were more elaborate than teletext and will be discussed later.

The BBC originally described CEEFAX as a "magazine" that was delivered electronically to the home. That was when it consisted of only 32 pages. Later it was increased to 100 and eventually to 400 pages. Considering the fact that if a page of print is to be legible on television it can carry at most only one-fifth as many characters as a printed page, a 400-page teletext edition is equivalent only to an 80-page published magazine. So probably "magazine" is still the best descriptor to use; it could hardly be likened to such familiar information sources as the typical 1,000-page household almanac or the 28,000-page home encyclopedia.

The term "magazine" also implies that it is a periodical, but teletext differs markedly from most published magazines in that it can be updated constantly, hourly if need be, and therefore can carry news, weather data, stock reports, emergency information, and the like that even the daily newspaper cannot adequately provide. In comparison to radio or TV newscasts it is superior because it is always available, literally within a few seconds. It is not necessary to wait until a scheduled broadcast or, once tuned to a broadcast, to wait for the particular information of interest to be given. With a teletext system one simply switches to teletext, waits a few seconds for the index page to appear, and then selects the specific information desired by punching the appropriate page number. It is this selection of particular pages by the user that has led to these systems being called interactive.

Since the BBC operates two channels, BBC 1 and BBC 2, it is able to present two CEEFAX magazines of the same length. With the addition of the IBA ORACLE service, there are three magazines already available to the properly equipped British viewer, each with 400 pages or so, and somewhat different content. Assuming that these systems coordinate their programming and don't duplicate each other too extensively, the home viewer has around 1000 pages of constantly updated information immediately available. Of course, the information is of general interest — for example, weather information, news summaries, football scores, and financial news — all in the nature of summaries or highlights because of the limited number of pages.

There is also the possibility of offering different magazines for different times of day: a morning edition, for example, with women's features and up-to-the-minute financial data, then an after-business-hours "evening edition" containing, say, vacation and entertainment information not presented during the day. Weekend editions could conceivably be different from weekday offerings. Already the BBC is providing different magazines for different parts of the country. CEEFAX is a BBC broadcast service, for which no additional receiver tax is required. IBA's ORACLE is indirectly advertiser supported, since it is from advertising that the IBA receives its income.

Reading from a TV screen is different from reading a book or magazine held in the hand. The description that follows here applies not only to teletext, but to any system that uses the home receiver for display, including, of course, broadcast television. The main difference is that the screen is effectively more distant and smaller in the eye than a page of print.

With a typical 18-inch screen, most people like to sit about 9 feet away from the TV set, a distance of about 6 screen widths. They prefer not to sit very close because the definition of the usual TV picture is relatively poor compared to home movies, or especially 35 mm slides, and the screen is uncomfortably fuzzy close up. At 6 screen widths the screen subtends an angle of 9 degrees at the eye, a very narrow angle, about the same as a 2-inch column in a newspaper held at reading distance. If you doubt this, just hold a page of newspaper up in front of you, at reading distance, so you can see the TV screen past it, and compare the apparent width of screen and column of type. The result of these conditions, the relatively small screen and the relatively great distance from which it is watched, is that you can't expect a line of more than 35 or 40 characters (including character spaces) to be legible, simply because smaller letters would exceed the limits of visual acuity.

CEEFAX and ORACLE put 40 characters into a line of print on the TV screen. A typical newspaper column will have between 30 and 40. Television producers have long known that the average typewritten page line of 72 characters cannot be shown on TV, and that in order for a page of typed print to be legible when shown in its entirety, it cannot exceed half a typewritten line in width. CEEFAX'S 24 lines of 40 characters per line constitute a page containing a maximum of 960 characters. (There are over 7000 characters on a full almanac page and around 8600 on a typical page of an encyclopedia.)

Computer terminals with CRTs (cathode ray tube displays) are something different. Such screens can legibly display even more characters than a typewritten line, up to 80 or, in some cases, 132. Of course, this would be below the limit of visual acuity if seen from a distance of 6 screen widths, but the computer user sits about 1.5-2 feet from the display screen. This would be some 2 or 3 screen widths distant, where far greater detail is visible. No home TV set would stand this close an inspection, so a higher definition black-and-white tube is used, and a different method of scanning than standard video is employed. The standard television screen thus has its limitations, but it also has one tremendous advantage: Nearly everybody has one.

The technical method of sending teletext, piggyback, as it were, on the regular TV broadcast, is interesting to understand. Explained very simply, a method was found to send a whole row of characters on those two otherwise unused lines between every field.[1] (Letters that size would require at least 25 lines of ordinary video.) The trick is not to try to send the size and shape of the symbol, but simply to represent the symbol by a set of pulses; seven bits is enough, using the binary counting system, to represent 128 different things, enough for all the letters in any European language in both upper and lower case, all the numerals, and then some. Since only the code is transmitted, not the character, very little bandwidth is required. In the receiver a character generator reads the code and generates the necessary video signal then and there, to display the character in as many lines as necessary on the standard TV screen. This information is recorded so that it can continue to be played back to the TV screen, just as any video signal, 25 times a second for as long as the viewer requires. Because the image is the same each time it is displayed, rather than changing, as motion video would, it appears as a still image.

With all of these complex functions necessary, it is clear that an adapter, converter, or a special TV receiver with this additional equipment built in would be required. In Britain, where the usual method of acquiring a TV in the home is by renting, a TV set equipped or adapted for teletext rents for a little over 25 percent more than the equivalent nonteletext set.

All 400 pages of CEEFAX are not recorded at once, however, another device known as a "frame grabber" selects only the page that has been requested, hence the short wait for access after punching up the page number. When another page number is punched up, the frame grabber starts watching for it. When it is found, the memory is erased and the new frame is recorded. All 400 pages of CEEFAX "go by" in about a minute and a half, which would be the maximum wait for any given frame if the user happened to call for it just as it passed, and had to wait through the entire next cycle. Mean access time is just half of that, or 45 seconds.

When CEEFAX consisted of 32 pages, average access time was only 3 or 4 seconds; with 100 pages the time increased proportionately to over 12 seconds. The decision to quadruple the number of pages also quadripled the average access time, a tradeoff that was evidently considered reasonable for the advantage of the extra pages. The waiting time was then comparable to the time a motorist is held up at an average stop light.

Newsflash

One of the features of CEEFAX is the Newsflash page, a means of alerting users, while they are watching television on their receivers, when an important news development has been added to the news highlights section. To utilize this capability, a user simply puts the system into superimpose mode, which allows him or her to both watch a TV program and simultaneously receive CEEFAX information, and selects the Newsflash page. The user may then proceed with TV viewing without interference from the Newsflash page, since it is normally blank. When a Newsflash is added to the system, a headline will appear on the screen and the user may then call up the appropriate news page for details.

Teletext Graphics

Teletext cannot be said to have good graphic capability. Because it is a character-based system, a kind of crude graphic is possible, built of blocks and straight corners; the individual "building blocks" are smaller than a character. This is quite useful for decorative headings, underlining, large block letters, and the like. Even stylized but recognizable maps are possible. This is similar in some ways to the graphics that are programmed into computers for a tour de force, composed of nothing but alphameric characters and printed out on wide computer paper.

The smallest possible picture element in the computer map is the character, resulting in a rather coarse matrix. A printed-out portrait of John F. Kennedy, for instance, is at least 18 inches high and must be viewed from across a large room to look realistic. Teletext could do better than this; it was possible to break each character space up into a matrix of 6 smaller areas, 2 wide and 3 high. (This is the most complex matrix that can be represented by 7 bits; the next larger would be 3 by 3 and would need nine bits.) Within this 6-cell matrix, 60 different patterns could be formed from varying combinations of white and black cells. Each of these 60 patterns then became new "mosaic characters" (see Figure 19). By giving many of the 128 possible 7-bit codes two different meanings, just as a shift lever will make a typewriter key type 4 in one position and $ in the other, 224 things could be encoded for transmission. Many of these were instructions, such as which of six colors or white an alpha character or a mosaic character should be, whether it should flash or remain steady, be displayed normal or double height, and the like.

ANTIOPE

The French ANTIOPE system is classified as teletext but has much more flexibility than the British systems. There is also a version of ANTIOPE (called TITAN) that is fed over telephone lines and hence classifies as videotex. ANTIOPE-BOURSE, one of the teletext services, supplies 80 pages of stock quotations, updated by the minute, with a maximum potential capacity of 400. Stockbrokers receive it in their offices off the air.

Rather than being limited to 400 pages, using only 4 lines per frame as the British systems do, ANTIOPE can be immediately adjusted to use any number of lines, up to the full channel bandwidth if desired. It is conceivable that at certain times of day television broadcasts, such as news or talk shows, could satisfactorily operate using a screen area of less height and hence fewer lines. There would be a blanked-out area across the bottom, for example, the scanning lines of which would be used for the transmission of a greatly extended teletext service. After broadcast hours the entire channel could be used to transmit ANTIOPE teletext, providing a service, theoretically, of 100 to 150 times as many pages as CEEFAX, continually recycled in the same manner.

Reuters IDR

The proprietary system known as IDR was developed in the United States by an American subsidiary of Reuters Ltd. IDR stands for Information Dissemination and Retrieval. This system is similar to teletext except that it utilizes more than 4 lines of the television signal. In fact, IDR utilizes an entire television channel. This, in the United States, consists of over 500 scanning lines per frame. Compared to CEEFAX's 4, the difference is about 125 to 1. It can be expected that for the same average access time IDR's potential capacity would be greater by a similar factor. However, IDR carries only 20 times as many pages as CEEFAX.

The IDR commercial and home services differ in format. Investment firms display the data on high-definition CRT screens similar to those used in computer terminals, which can display 64 characters per line. The IDR home service, which must be viewed on lower-definition home TV receivers, provides only 32 characters per line. In either case, however, there are only 16 lines per page, compared to CEEFAX's 24. This largely reflects the reduced detail generally visible on 525-line U.S. standard TV, as opposed to the 625-line European standard used in Britain.

IDR, like CEEFAX, transmits its entire library of information constantly; row-grabber and storage equipment at the receiving end pick off each item of information as it is requested, and record and display it.[2] Total recycle time is about 5 seconds, reducing mean access time to 2.5 seconds. This is one-twentieth of the waiting time encountered by the users of CEEFAX. Had Reuters felt that a longer access time was tolerable, IDR could theoretically have had a proportionately greater capacity. Still, the total store of information is very great; 9000 pages, compared to CEEFAX's 400.

Reuters has in the past provided various alphameric services via telephone lines to the financial communities in London, New York,

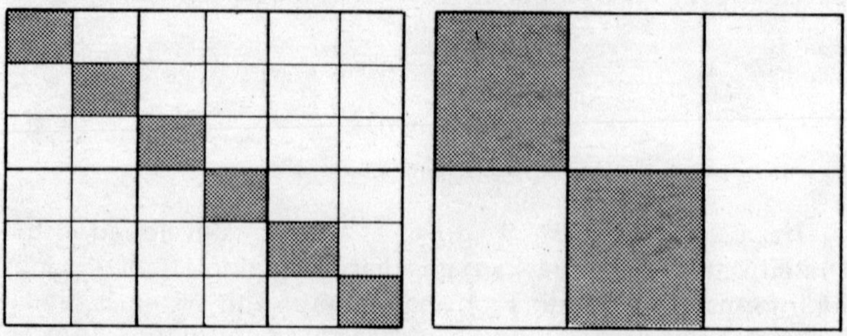

a. Six continuous mosaic characters representing a diagonal line. On the right, for contrast, is the closest one could come to a diagonal line using whole characters of black or white instead of the six part mosaic characters. In encoding only six bits were used to represent the six areas of the mosaic characters; the seventh was left as a zero for future need. Thus the first character on the left is represented by the binary code 1001000, the second by 0000100, the third and fourth by all zeros, and so forth.

Figure 19: Comparison of alpha-mosaic and alpha-geometric graphics.

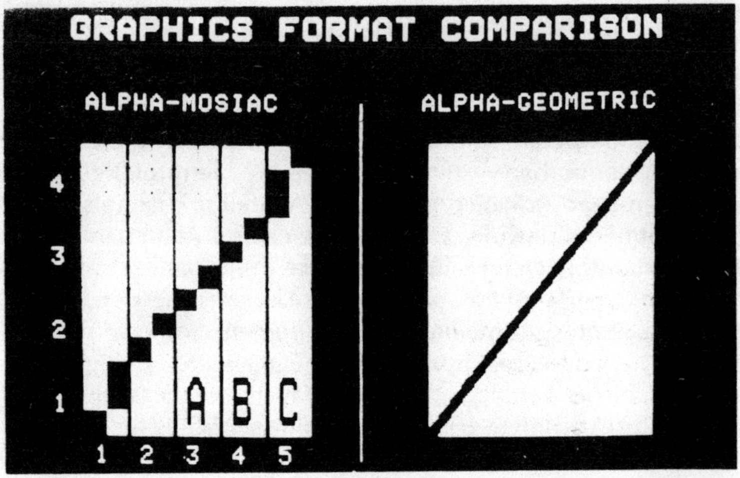

b. A further comparison, involving, in this case, 20 mosaic characters in the alpha-mosaic example on the left representing a diagonal line, and a similar picture area with a diagonal line as represented by the more sophisticated alpha-geometric method.

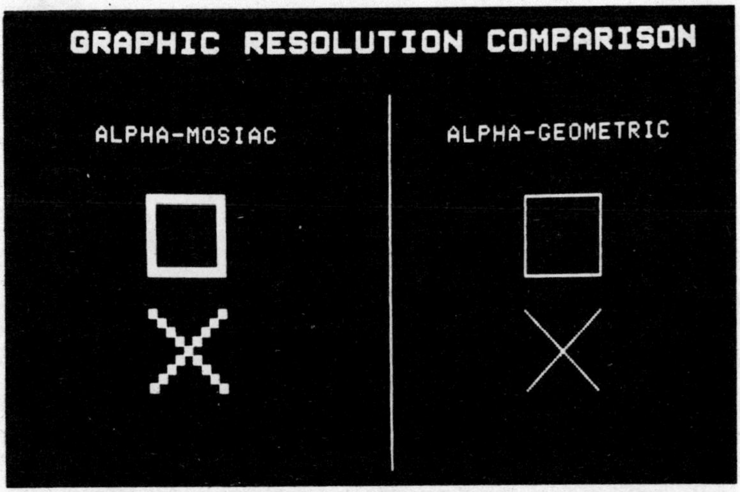

c. A comparison of the amount of detail possible in these two methods of representing graphic images. ("Resolution" is an engineer's term roughly synonymous with "definition.")

Figure 19: (Continued)

Toronto, and other cities, providing up-to-the-minute information on securities, commodities, money rates, and market quotations, all items dated and timed. The IDR home service is directed to a broader public and hence carries various other kinds of information, such as weather, sports, horoscopes, and the like. Since it requires a full TV channel, and broadcast channels are scarce and expensive to operate, it is disseminated via cable, where, at least on the modern urban systems, channels are plentiful. Reuters has taken a ten-year lease on one of the RCA satellite channels in order to distribute nationwide, and will have cable TV companies make an extra monthly charge for this service in the same manner that they do for pay TV.[3] At the start the entire business service, containing some 80 percent business and 20 percent home service content, will be fed to cable homes. Without payment of a higher rate, however, the householder cannot display the business information — he or she can, however watch the home service, which is carried on the same channel. By address code Reuters can tell terminals what they can and cannot receive.

So far IDR has been strictly alphameric; no graphic capability, even as crude as that of CEEFAX, is described in any of the Reuters promotional literature. However, the IDR home service will have character-based graphics similar to the European teletext systems, and will be displayed in six colors and white. Probably Reuters considers that its commercial audience of bankers, brokers, and the like is primarily interested in information content in which character-for-character printed words are far richer than mosaic character graphics (alpha character for mosaic character). Financial information is best expressed in tables of figures; maps, diagrams, and graphic representations of statistical concepts, at least for the professionals, take a back seat to the more direct, and compact, presentation of the data themselves.

Most users of the commercial IDR system want to call up specific stock and commodity reports, for which they need full alphameric keyboards. However, Reuters terminals do not use the standard QWERTY typewriter format. This is because most brokers and other users are hunt-and-peck typists, and for them, the letters are much easier to find when placed in alphabetical order.

Reuters also offers a service via telephone lines called MAT (Multiple Access Terminal) for areas where there is not yet access to a cable TV system carrying the IDR service. MAT, however, is an example of a videotex system, several of which will be described in Chapter 8.

notes

1. The standard American TV signal contains 525 lines per frame, 30 frames per second. Since there are two *fields* per frame, each field consists of 262.5 lines, and there are 60 fields per second. The lines are interlaced — odd-numbered lines in the first field, even-numbered lines in the next — to reduce flicker.

2. Instead of grabbing total frames, IDR grabs what they call "envelopes" of data —1 per line in 32-character format, 2 per line in 64-character format — thus the term "rowgrabbing" rather than "frame-grabbing"

3. RCA's Satcom I, II, and III and Western Union's Westar I and II plus other U.S. and Canadian domestic satellites hover about 22,300 miles above the equator over the eastern Pacific Ocean, directly south of the North American continent. Special narrow beams serve Alaska and Hawaii.

8

VIDEOTEX

The information system known generically as "videotex" was at the time of this writing under development in at least nine countries. It was in actual operation only in England, however, where it was pilot tested under the name Viewdata. The service that was based on this technology and later introduced in the market trial was called PRESTEL. Videotex systems share one important thing with teletext systems — both use adapted home TV receivers for display. Videotex differs, however, in that it is not broadcast or cablecast as part of a regular TV signal, but is fed into the home by wire, generally using standard home telephone lines. Television is not involved, except that the TV receiver is used for display. Furthermore, it does not send its entire store of information every few seconds and then recycle as the previously described systems do. Only those pages that are requested are actually transmitted to the user's home (or office). It does tie up the phone when it is in use, so it is presumed that heavy users may want to have a special line installed for the purpose.

PRESTEL

PRESTEL was developed by the British Post Office as a new telephone service. It takes advantage of the fact that each telephone subscriber has a private line all the way back to the central exchange (excluding party lines, of course). It is not like cable TV, where all subscribers to the basic service are connected to the same cable and receive the same programs. Page selection is thus done remotely, by upstream signals from the home, and the switching actually takes place at the central exchange. This then requires a two-way transmission system — one channel in each direction — which of course already exists in the telephone system anyway. Teletext and IDR, on the other hand, are both strictly one-way systems — everything is sent and the user selects what he or she wants to read; no signals flow upstream.

Because viewdata systems do not transmit and constantly recycle their total data base, they have no theoretical limit to capacity. The PRESTEL system began with 50,000 pages and expanded quickly to 100,000, which was all that the computer could handle. With another computer this upper limit was nearly doubled. By 1980 it was being expanded to allow for a 500,000-page potential size.

Likewise, there is no theoretical limit on the number of users that can be served simultaneously. The British Post Office talks easily about serving all television homes in the country. When the number of users exceeds 200 per computer, however, waiting time exceeds a few seconds and begins to be noticeable. Delays of up to 30 seconds or more are familiar to pioneer computer users who share time on large central computers, but such delays might not be tolerable to people not as ready to make allowances. PRESTEL's answer will be to add computers and keep the access time within tolerable limits.

The fact that viewdata systems such as PRESTEL are not limited in capacity, as other teletext system such as IDR are, justifies our thinking and talking about them as "electronic newspapers." Certainly the content of all these information systems, being highly timely and topical, is more akin to the newspaper than to the magazine.

The amount of information that must be typed into digital form for insertion into the files of PRESTEL is tremendous, and so are the person-hours required for this as well as for the task of constantly updating and revising. With a data base 500 to 1,000 times as great as that of CEEFAX, for instance, PRESTEL happily does not attempt to edit and insert its own material. Instead it relies on information providers (IPs) — organizations small and large, commercial and noncommercial, that have a need to disseminate information, either to announce a service or as a service in itself. PRESTEL, like other services of the post office, is a common carrier, available for anyone to use who will pay the cost. Pages in PRESTEL are leased by the year. There is a minimum of 100 pages required per supplier; some organizations have provided up to 10,000 each.

Some information providers are large firms, such as banks and tourist agencies, whose pages contain information on the services they have to offer, and hence are generally free to the user (except for the telephone line and PRESTEL cost component). Other free (or low-cost) pages are supplied by government agencies whose services to the public are normally free and who are using PRESTEL as one of many ways to deliver these services. Examples are the British Tourist Authority, the Meteorological Office, and the Department of the Environment. Many suppliers are in the information business directly and use PRESTEL as a means of bringing their products to market. They may charge the user anything from half a penny per page (the usual cost) to as much as half a pound. Many of these are newspaper and other publishers who already have large data bases with large staffs constantly involved in updating. PRESTEL offers them an immediate means of cashing in on their services. Included in this category also are suppliers of detailed financial data and consumer reports. PRESTEL began operations in 1978 with about 150 information suppliers.

Information suppliers must have more elaborate terminals than home users since they require a full alphameric keyboard plus special keys for making graphics, selecting colors, and the like. This equipment is hired from the post office. The kind of graphic effects that are possible with PRESTEL are the same as those described for CEEFAX and ORACLE. The smallest picture element (pixel) is one-sixth the size of a character, so there are 5760 pixels in the picture area. Compared to European standard broadcast TV, which, when operating at optimum quality (over 600 scanning lines, 500 variations per line), presents an image with the equivalent of some 300,000 pixels, realism and fidelity of the PRESTEL graphics is very low.

In order for PRESTEL to be self-supporting, the user must be charged a sufficient fee per page to defray three kinds of costs: (1) the costs of writing, editing, and inserting information into the system, (2) the costs of the use of the telephone line while viewing PRESTEL, and (3) the post office's handling charge for PRESTEL. The user may or may not be charged for the first item, depending on whether this revenue is coming from advertisers, organizational budgets, or the users. The phone call cost will always be incurred, at the same rate as the subscriber normally pays for local calls, and the final post office PRESTEL handling charge, which amounts to only half a penny per page, is likewise inescapable.

Section a of Figure 20 diagrams a typical system for display of alphameric and graphic material on the home TV screen. It can represent either teletext or videotex, the main difference being the transmission system, which in the case of teletext is a TV channel, or an unused portion thereof, and in the case of videotex is the telephone line, which of course is two-way.

Section b of Figure 20 applies the diagrammatic model to teletext. Note that the system is one-way; the user's keypad makes the selection within the user's own terminal, which in the case of teletext contains a frame-grabber to obtain each selected frame as it cycles by. Only two parties are involved in the teletext system as it now operates: the broadcasting authority and the user.

In section c of Figure 20, a videotex system is diagramed, showing transmission via a regular telephone line, which carries the user's keyed instructions back to the central store where the frame selection is made. Since the IP and the central store are usually in different locations, a phone transmission link between them is indicated. (In the case of teletext they are probably at least in the same building.) Note that there are three parties involved in this system, which is exemplified by the British PRESTEL, the only operational videotex system at the time of this writing. Sections d through h of Figure 20 diagram five possible variations on the videotex model, characterized by different combinations of parties and their functions.

Section d of Figure 20 shows the "umbrella provider" system. Since a minimum booking of 100 pages is required of any IP by PRESTEL, at a minimum cost of 400 pounds per year, and an IP terminal must be rented from the post office, smaller organizations that would like to book only a few pages cannot afford to do so. The umbrella provider books a large block of pages, which are then retailed in small amounts to the small IPs. The umbrella provider rents and houses the terminal, which the smaller IPs pay for on a pro rata basis. There are thus four parties involved in this variation of videotex.

In section e of Figure 20, the middleman, who we have just seen as an umbrella provider, might take over the central store as well as

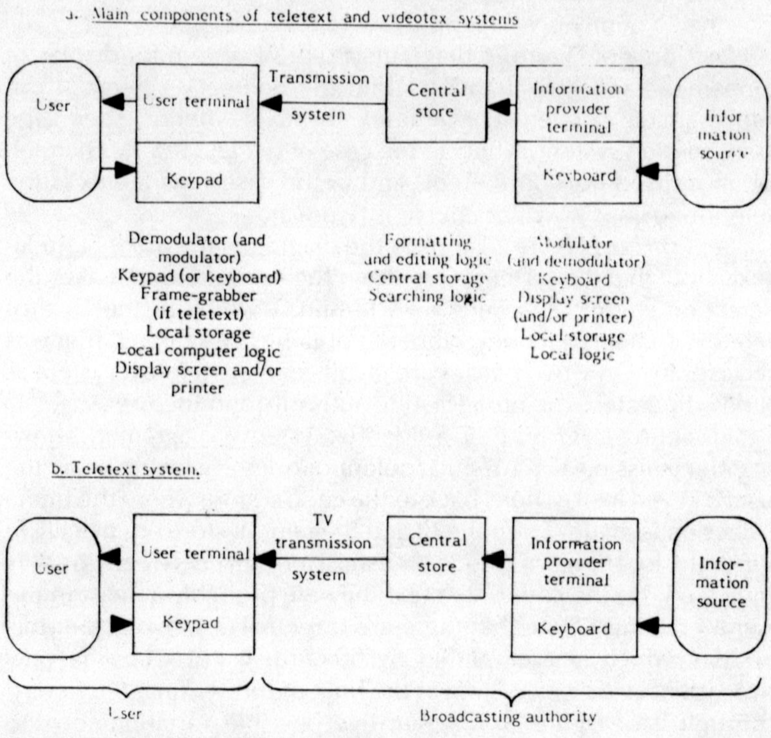

Figure 20: Teletext, videotex, and variations.

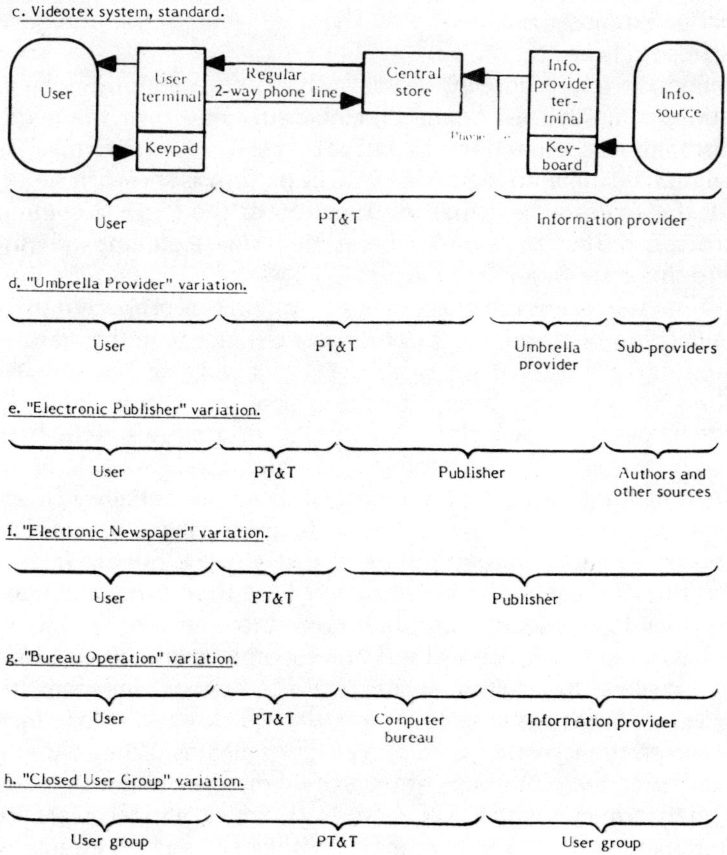

Figure 20: (Continued)

the IP terminal. The function of this position then becomes broad and important, analogous to that of publisher in the world of print. IPs would relate to the holder of this position in much the same manner that authors and data firms relate to the print publisher. In the electronic publisher application there are again four parties involved.

Section f of Figure 20 diagrams the electronic newspaper. Like a newspaper publisher, the electronic publisher hires a staff of reporters and writers and builds up a data base. This approach further extends the role of publisher and reduces the number of parties to three.

Section g of Figure 20 shows still another pattern in which a middleman, such as a computer bureau, takes over the central store and such functions as PRESTEL itself now performs. This variation is similar to the basic PRESTEL pattern as shown in section c of the figure except that the function of the PT&T, a common carrier, is reduced to simple operation of the telephone network. Here there are four parties again.

Finally, in section h of Figure 20, a system is represented that is actually in operation in London, in coexistence with the standard three-party mode of using PRESTEL. Called the "closed user group," this system has only two organizational components: the British Post Office, which manages the central store and the transmission system, and the user group, which contains both IPs and users, in some cases interchangeably. This is essentially a closed-circuit use of the videotex system. The information inserted into the system and recorded in the central store is not available to unauthorized users who are not part of the user group and hence not privy to the secret code that gives them access. A company such as a bank, with many local offices, an investment firm, or an auto dealers' association, could use the system whenever it is necessary to make the same information quickly available to many dispersed branch offices, employees, or affiliates. When videotex is used for intercommunication in this manner, many users will need IP terminals with full keyboards. If senders and receivers thus interchange roles, videotex can be recognized as a fully interactive system in which two or more communicants may respond to each other.

A user group could also be a set of hobby enthusiasts, railroad fans, for example, although the cost of rental or purchase of IP terminals might discourage this kind of use for at least a few years. While at the onset this "closed user group" application of PRESTEL may be minor in comparison with the one-way applications that tend to follow the broadcasting model, it could easily turn out to be the most important and most lucrative application in the end. It is a first step toward computer teleconferencing and computer mail in the home.

The viewdata videotex technology that underlies the PRESTEL service has been directly purchased by the PT&Ts of several other countries, since it was the only system fully up and running, with most of its initial problems solved. In most cases the technology was adapted to the particular needs of the country. Viewdata will be a government service in West Germany (where it will be called *Bildshirmtext).* By 1980 the Netherlands, Switzerland, Hong Kong, and a Dutch publishing company had also purchased PRESTEL.

Titan

The videotex version of the French ANTIOPE system (TITAN) will, according to the reported plans of the French PT&T, launch into a nationwide service that would put it at the very core of intercommunication in that country. It is proposed that all 20 million subscribers to the French telephone service be provided, at government expense, with small-screen TITAN terminals. Then the complete French telephone directory will be made available as a data base. There will be no hard-copy printed telephone book produced. It is expected that the savings resulting from not having to fund this gigantic regular printing will pay for the costs of conversion. Other data bases will be added later as various information suppliers enter the scheme.

Telidon

The most interesting videotex system so far developed was born in Canada, a child of the Canadian Department of Communications. In a country that spans seven zones from east to west and in which the largely wilderness area extends from the latitude of Rome to that of northern Greenland, it is logical that communication should be an important government concern.

Canada was not satisfied with the design of videotex systems as developed in Europe for at least two reasons: (1) The definition of graphics was far too crude to convey any real information and (2) the systems were designed for, and locked into, today's limited-quality broadcast TV receivers. Once the European alpha-mosaic (character-based) graphics technique is accepted, reasoned the Canadians, "it will be impossible to send higher resolution images without knowing the display characteristics of the receiving terminal" (Brown et al., 1978: 2). They could foresee a videotex central store having to provide multiple versions of each image to suit a variety of new and higher-quality display systems. In anticipation of this need, they preferred a system design that could transmit images in a manner independent of terminal design, on which there could be no foreseeable constraints. This meant abandoning the simple alpha-mosaic graphics process and developing a system that has come to be called the alpha-geometric.

Furthermore, the Canadians reasoned that they would probably in the future want to extend their essentially one-way videotex system into a fully interactive medium through which one subscriber could transmit visual images to another subscriber, or engage in interactive games or cooperative work projects. Widely separated architects, for example, could each draw, and erase, on the same visual area. In short, they could foresee the time when a system would be needed by means of which any member of the public could become an information provider, inserting his or her own alphameric and graphic information into a central store.

A system for describing and transmitting graphics was then developed that the Canadians felt "must be considered as the only sensible way of implementing videotex systems which can grow and be adopted without constraint, as technological developments bring complex equipment within the buying range of the average consumer" (Brown et al., 1978: 3).

The method of converting graphic information into digital (numeric) terms so that it can be stored and accessed is called "description." Teletext (and PRESTEL) graphics, as we have already explained, are based on mosaics of six smaller areas (see Figure 19, Chapter 7). The smallest possible pixel is thus one-sixth the size of a character. The total number of pixels in the entire frame is 6 times the maximum number of characters; in the case of PRESTEL, 40 characters per row times 24 rows times 6, equal to 5760 picture elements.

In the Telidon system, which does not use mosaic characters, the definition of the display depends on the capability of the terminal equipment being used. A character-oriented terminal, for example, may display as few as 4,800 pixels and a home TV receiver 76,800 pixels, while a high-resolution business terminal can be capable of 1,228,800 picture elements, all from the same Telidon store. This is a range of over 250 to 1 (see Figure 21 for a depiction of Telidon alphamerics).

Technically, the alpha-geometric method of graphic description is based on the use of PDIs (picture description instructions). Transmitted codes, instead of representing mosaic characters, represent picture instructions. This is almost analogous to the communication game in which one person, limited to the use of words alone, tells a group how to draw a simple diagram or picture, after which the various versions are compared to see how well the person communicated.

There are seven basic directions that tell the drawing beam what to draw — such as "point," "line," "arc," "area," and "polygon," — plus instructions on color, gray scale, and areas to be filled in with a choice of four different textures. Just as in the case of CEEFAX the code "words" contain only eight bits; seven are used and one is in reserve for unknown future need. A "shift" code informs the receiving terminals when to interpret the codes as alphameric characters and when as PDIs. As with any videotex system, an adapter unit is necessary at the receiving terminal. In Telidon this records the signals transmitted, interprets the PDI command codes, and, in playing back the recording, keeps refreshing the displayed image every one-thirtieth of a second until a new image is ordered (see Figure 22 for an example of a Telidon map).

Combination Media

Because CEEFAX and PRESTEL are compatible, British viewers with PRESTEL receiving equipment are also able to display CEEFAX

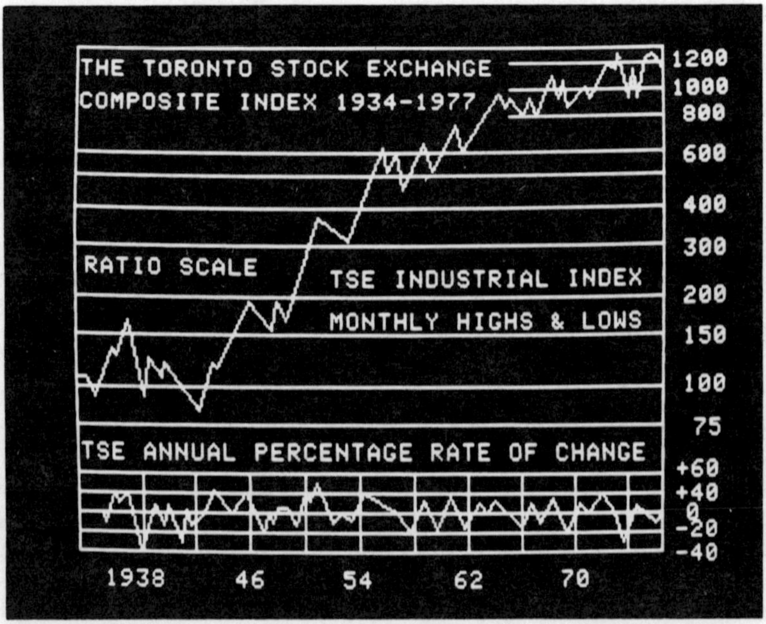

NOTE: Courtesy of the Government of Canada, Department of Communications.

Figure 21: Example of Telidon alphamerics.

on their receivers. In fact, the PRESTEL keypad, which contains some thirty buttons, includes controls for operating CEEFAX as well. A broadcasting organization, KSL-TV in Salt Lake City, Utah, has developed a system that augments a teletext system in a limited manner, basically relying on teletext but accessing a few pages by telephone, as needed, to extend the capacity of the system. KSL-TV teletext incorporates 100 pages on a 7-second cycle. Beyond this, a system that they call "touch-tone teletext" allows the viewer to access various additional decision pages and finally the information that he or she wants.

A telephone number appears on a teletext page, which, when dialed, is answered by the teletext computer. A special decision page is then flashed to the user, on a once-only basis, listing the various options that he or she may select using touch-tone buttons, thus making it possible for page after page to be flashed to the user in the same manner. Meanwhile, a large number of other users

NOTE: Courtesy of the Government of Canada, Department of Communications.

Figure 22: The alpha-mosaic graphic system compared with the alpha-geometric, in representing a map of Canada.

may be doing the same thing using the same teletext page. If 100 people were to access via telephone simultaneously, at one-thirtieth of a second each, it would cause a little over 3 seconds delay to any one user.

Note that only selection is done via the telephone; the selected pages are sent to the viewer as teletext, so the system cannot be called videotex. This means, of course, that these pages are also sent to everyone else that happens to be using teletext at the time. This has no effect, however, on any of the other users except to briefly reduce the full number of pages and, since each page is sent only once, this use lasts only a split second (one-thirtieth of a second in the United States).

Recording

The value of hard-copy printouts of informational materials for filing and future study is attested to by the heavy use of photocopiers in libraries. Reuters has for some years supplied its financial customers with teleprinters on which at least the alphameric portion of teletext can be printed out, either while being displayed on the TV screen or without screen display. The ITT company in England offers a small facsimile printer that turns out 5-by-7-inch pages that record all alphameric and graphic components of the teletext signal. How practical this can be, of course, depends on its cost.

The teletext signal may also be recorded on audiotape cassette and redisplayed when needed. The problems of indexing an audio cassette would probably make retrieval of such recording rather cumbersome, however.

SYSTEM DESIGN FEATURES: ISSUES AND PROBLEMS

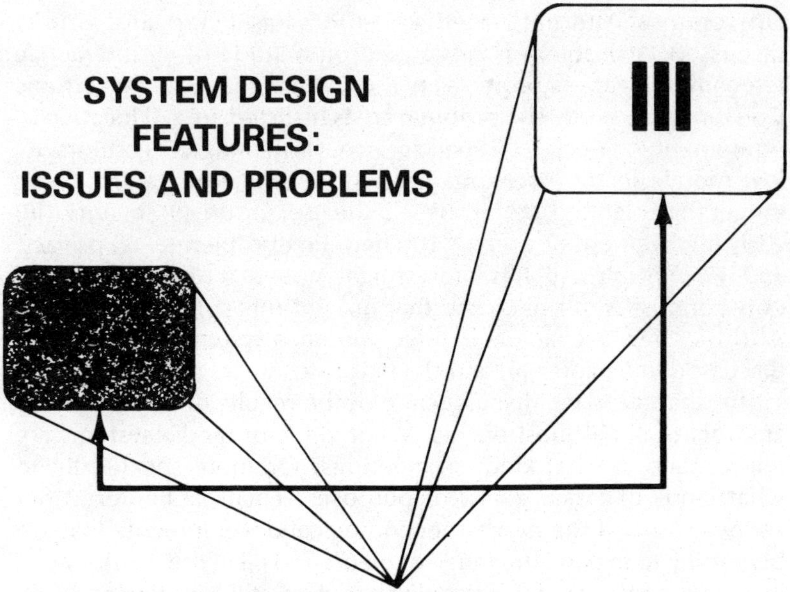

The characteristics of interactive systems that affect their success or failure are of two kinds: those intrinsic to the system, and those that are extrinsic. Both must be considered in designing a system and they often interact, but, for purposes of analysis and discussion, it will help if we separate them into these two categories. *Intrinsic* factors are the technical and human factors of the system itself, those characteristics of interactive communication systems that have been the major concern of this book. *Extrinsic* factors arise from the situation in which a system is used. These are mostly human factors; there are many of them, and they have probably the greatest influence on whether a system is successful or unsuccessful in operation.

Failure to recognize the difference between intrinsic and extrinsic factors and to determine which are at fault has led many organizations to abandon systems, branding them unworkable, when they were merely unworkable in the particular situation, under the particular extrinsic factors that obtained. To give a simple example of this, financial resources are extrinsic to a system. Cost is intrinsic. A high-cost system can be workable in a situation that provides sources of copious operating funds and unworkable where only limited resources are available. John Carey of New York

University's Alternate Media Center, has listed and briefly discussed 13 problem issues based on a study of 45 interactive telecommunication projects in the United States (in Electronic Conventions, 1979). The problem areas he listed are: (1) location of equipment, (2) need, (3) repair service, (4) training, (5) comfort, (6) low morale in the user organization, (7) implementation taking longer than planned, (8) turnover in the user group, (9) security, (10) early involvement of users, (11) scheduling problems, (12) privacy, and (13) multiple use. It is interesting to note that of these 13 items, only comfort is intrinsic, and that not entirely so; the rest are all extrinsic. Intrinsic factors in teleconference system design will be discussed in Chapter 10, and the extrinsic factors in Chapter 11.

In Chapter 12 we discuss some of the results of the search for answers to such questions as: What kinds of media systems are appropriate to what kinds of meetings? Or, more specifically, to what kinds of tasks, goals, or purposes? Finally, Chapter 13 describes some of the newest technological developments that are beginning to expand the range of media and fill in the blank spaces in media continua, for example, between still pictures and full-motion video.

In the first chapter of this section, however, we must examine an area that contains many factors that can make a big difference in the effectiveness of a system. It is often called simply the area of "how" one uses a system. In a sense it is more basic than the medium and any of its characteristics. It is more "content" than "medium" — it is really concerned with how the "means" of communication are used. It is the realm of technique.

9 TECHNIQUE: THE USE OF THE MEANS

The reader will recall that in the introduction to this volume we introduced a theory of communication in which we identified "means of communication" as something separate from media. Natural means that we are born with, such as vocalization, gesture, facial expression, and bodily movement and position are capabilities we share with the higher beasts. These were contrasted with artificial methods, such as spoken and written *language,* drawing, gesture *language,* and the like, which are devised by human beings and must be taught and learned. Together these are the basic means of communication used both in face-to-face communication and via media. A person who is skilled and effective in communicating face to face is thus called upon to use the same skills and techniques when communicating via media. Such a person can be expected to be equally effective in media, provided two things: (1) that he or she can overcome the lack of feedback from listeners that is characteristic of one-way media, and (2) that he or she can accept the limited range of means that all media impose to one degree or another. In radio, for instance, one may speak (or have others do so), sing, make music or other sounds, but there one's means end. This can also be a blessing in disguise. It was

Goethe who said that the master first reveals himself when working within limitations.

"How" one uses radio, television, or any other medium is a matter of production and performance technique — that is, how one sings, speaks, gestures, and the like — and the same skills to a large extent carry over into interactive communication. Particularly the conversation techniques used in face-to-face interaction transfer readily to two-way media, and a good conversationalist in one will tend to be good in the other also.

Each of these artificial means of expression, even gesture language, is institutionalized in our society. Teaching each of them has become a profession. One can go to language school, art school, dance school, singing school, and acting school; one can study writing, public speaking, mime, theater arts, film, and television, in which various combinations of means are used. While fine art has only a minor role to play in interactive communication, many performance skills are indeed directed toward developing and maintaining a relationship between a performer and his or her face-to-face audience, and these can in a true way be considered skills of interaction. A person who wants to become a media performer or teacher must first learn performing or teaching and the appropriate skills of communication — that is, how to use the means of communication. Most people have their greatest opportunity to learn and practice these skills in face-to-face situations.

This is a book on media of interactive communication, not means; the domain of means, basic as it is, falls outside the purview of this book. There are some good sources for understanding these means and how best to use them. The University of Wisconsin Extension (see references; Monson, 1978), through several publications, has very effectively organized and presented what they call "Four Steps to Interactive Teleconferencing." These are (1) humanizing, (2) participation, (3) message style, and (4) feedback.

"Humanizing" is personalizing your communication and creating group rapport. "Participation" includes a variety of techniques for activating the participants. "Message style" is how you present the information. "Feedback" includes techniques of collecting information during and after interactive sessions that will help you evaluate your success in achieving your objectives.

While reading the many excellent techniques and suggestions for effective use of interactive telephone for instruction presented in these guideboooks, it becomes obvious that they are not techniques of teleconferencing alone, but techniques of instructional communication (and instructional management) — in short good teaching. These books, especially the volume by Monson (1978), are highly recommended for anyone who is interested in how best to use audio, augmented audio, television, or any other medium based on the spoken word. Even if instruction is not your purpose, humanizing, participation, and even feedback are important to almost any teleconferencing purpose, and there is much in these guidebooks that is of value. They are also recommended to classroom teachers, who waste their golden opportunities if they do not make the most of their possibilities for interaction.

This is the most important concept to emphasize in this short discussion: In using an interactive medium, one must make the most of its interactive capability. However helpful it may be for a speaker to have two-way communication and be able to see and hear the responses of his or her audience, unless members of the audience also receive responses from the speaker, for them the situation might as well be one-way. It is difficult in a society in which the culture is based on one-way media — print, television, film, and radio — for people to use any of these media in such a new and different manner. The model of how to use television is before us all day long, and for most of us, has been for as long as we can remember. To have to ignore that model and turn instead to an informal approach, with people on screen behaving no differently than they do in everyday life, doesn't seem right. It doesn't seem like television.

And indeed it is not. Interactive communication is never fictional, rarely showmanlike or formal. One-way programming most often attempts to be. Even the category of programming known in Europe as actuality must have its formal program frame, its opening titles and music, its final fade out. Even the participants on TV game shows are thoroughly rehearsed and coached to give larger-than-life depictions of themselves — to sparkle, to laugh, and, if they win, to go really wild, jumping up and down, kissing, and hugging — assured that anything less will make a less exciting show.

In interactive communication, whatever the medium, there is never a show, for one very good reason: There is no audience. There are only participants. For whom would they be doing a show

—each other? The model for interactive communication is real life. For a two-way (or dyadic) interaction the model is the ordinary conversation — as might ensue between the same two people — not anyone else's conversation, but only theirs. The model for the teleconference is the ordinary meeting — as well planned and organized, controlled, and paced as the same people could do in a face-to-face situation, but not bascially different.

An example of this is how a teleconference begins. Some teleconferences begin neatly, cleanly, like a proper TV program. As the participants gather for the meeting an attractive slide or title card is on the screen, accompanied by soft music. At the scheduled time the music and the title fade out and the chairperson fades in. With perfect timing, the chairperson takes the cue, smiles and greets the "audience." It is acceptable to all because it fits the broadcast model.

More often, however, a teleconference begins like an ordinary meeting. The connection is made beforehand and is ready when the participants gather. You see a picture of an empty room, a table with a microphone or two, transmitted from the other end. Suddenly there is an off-screen noise, a door opening and closing. A shadow moves across the background and a man passes behind the table distributing papers at the various places. He peers at you to see if he recognizes anyone. He does see someone he knows, he smiles and says good morning, and someone sitting behind you responds. He says that one of the participants at his end of the line might be a little late but wanted the meeting to start without him. Finally the chairperson arrives and takes the central chair on the screen. He looks to see if all are present, greets people he knows, and asks if you are about ready to start. He also wants to know about the weather in your town and what the temperature was that morning. He introduces one or two of the participants at his table who may not be familiar to all and has you do the same. After some further chitchat he says something like, "Well, I guess we're all here — we may as well get started. Now, the first item on the agenda . . ."

This is a far cry from the attractive title card and music and the fade-out/fade-in. But a meeting has started, not a program; rapport has already been established, interaction has begun. Every bit of the seemingly useless interchange prior to the meeting, as in any meeting, makes the meeting easier, friendlier, more interactive.

People tend to act themselves in interactive communication, playing no role other than the one they play in everyday life or that they would play in the same meeting if it were held face to face. This can be amazing to a producer of broadcast TV who has had no experience with interactive television. One of a producer's greatest problems, when using nonprofessionals in a broadcast program, is to somehow get them to relax out of the stuffy, formal roles they want to play, to get them past their inhibitions and into acting themselves. They are stiff because they are not talking to anyone in the broadcast studio — they are called upon to imagine an audience at home to whom they must pretend to respond. In interactive TV they see the people they are talking to, see them smile, hear them respond. They can relate to someone and, relating, interact, and, interacting, be themselves.

10 INTRINSIC FACTORS IN TELECONFERENCE SYSTEM DESIGN

Many of the factors we will discuss in this chapter apply to all media based on the spoken word. Some will apply as well to alphameric media. Many, however, are related only to the visual component of audio-based media. As far as possible, we will try to keep the discussion clear by grouping the audio factors together and the visual factors together, and by explaining the more general issues at the beginning.

Goals

Two general goals seem to be pursued by teleconference system designers. These are (1) to simulate, within the limitations of the medium, the experience of face-to-face communication, and (2) to go beyond the possibilities of face-to-face meetings and develop new kinds or dimensions of communication. The first of these characterizes particularly the voice-based media, while the second is an objective mainly of computer designers. There are often other objectives as well, such as reducing the cost of communication. In the case of the common carriers, designing a system that will be as acceptable as possible to its potential clients, as quickly as possible, is undoubtedly a prime goal. While most of these goals are compatible, that may not always be the case. For instance, common-carrier-designed systems accept a requirement of privacy. These systems make a point about there being no operator present. *Because* there is no operator, however, users have to do their own switching, handling of graphics, lens zooming, and so on. This then conflicts with the goal of transparency, described above as the feeling of face-to-face communication.

In addition there may be subconscious goals, such as to simulate analogous one-way media, for instance, which was discussed earlier. Finally, we must realize that the design of a teleconference system is a joint operation requiring a team of specialists — engineers, industrial designers, human-factors planners, marketing specialists, researchers — and each of these may have a different set of subconscious goals, however well thay may agree on the goals they have been able to verbalize.

The two main goals — simulation of the face-to-face situation and extending communication beyond what is possible in the face-to-face mode — are not entirely incompatible, but sometimes seem to be pulling designers in opposite directions. The two main divisions of interactive communication — voice (or audio) based and alphameric based — have each apparently been aimed at a different goal. Voice communication is the most important method used in the face-to-face mode, so naturally the improvement of audio-based systems has been directed toward making interactive speech as "natural" as possible, which means as close as possible to the experience of face-to-face speech. Alphameric-based intercommunication has little hope of simulating face-to-face intercommunication, even when it is done in real time, so the thrust of development in this field is toward developing new means that face-to-face intercommunication cannot use.

The two approaches are complementary, in that one may be used in some circumstances for certain purposes, while the other is available for other circumstances. They are compatible, at least potentially, because media may be combined that have not been used together before, thus making it possible to convey means of communication in new and different combinations.

Simulation of Face-To-Face Communication

The closest simulation of the face-to-face situation results in the most transparent system. People sound and look as they would sound and look in the same room; users are least aware of the medium or the particular system being used. To those who believe that the closest possible simulation of face-to-face communication is the best possible design for teleconferencing, the following factors are generally considered ideals to approach as closely as possible.

Audio Factors

(1) Speaking must be normal. In a typical business conference people are in social distance (1.25 to 3.7 m) from one another and the teleconference system should require no less or no more voice volume.

(2) Listening must be normal. The *volume* of sound reaching the participants must not be more, or less, than the volume that would be experienced from participants within social distance. The *acoustics* must be normal — the combined room sounds and reverberation caused in the distant room before entering the microphone, and in the receiving room after leaving the loudspeaker, should not be noticeably different in this regard from sounds originating in the listener's room itself. The *quality* of sound must be normal. Fidelity must be realistic. There must be no interference from internal or external noise.

(3) Conscious operation of the equipment should not be required. Users should be unconscious of the *microphones* that pick up their voices. They should not have to remember to keep a certain distance from fixed microphones, or to hold hand microphones a certain distance from their mouths. Users should not have to adjust sound volume or make any other adjustment during a meeting. They should not have to turn microphones on or off or be careful in which direction they are pointed.

(4) Acoustic feedback should be no concern of the user. Feedback should not occur in normal operation and the user should not be burdened with responsibility for doing anything about it.

(5) Users should be able to speak at any time, just as they would in face-to-face interaction, subject, of course, to normal courtesy. This means that an ideal system should have the interrupt possiblity. This is generally achieved by open audio — all microphones open at all times. This means not only that people are able to interrupt each other at one location, but that a person in one location must be able to interupt a speaker at another location. This means that the audio must be open (or seem to be open) on a full network basis.

Already it can be seen that with the current state of the audio art, the full realization of one of these ideals may not be possible without conflicting somehow with another. Even full utilization of the audio technology that today exists may not be possible under typical cost constraints.

Video Factors

(1) A speaker must see at least some of the persons to whom he or she is speaking.

(2) Listeners must see the speaker.

(3)To a speaker, the apparent distance to the listener and to a listener, the apparent distance from the speaker must be appropriate to the kind of communication involved. That is, if it is a business conference, people on the TV screen should seem about as close as if they were across a conference table; for a personal interview or psychotherapy session, as close as across a desk or in an adjacent chair.

(4) Speakers should look as they would if they were face-to-face with the listeners. That is, elements of lighting, tonal values, color balance, and the like should be sufficiently realistic to be unnoticeable.

(5) Speakers should not have to be concerned about the cameras that are covering them; they should not have to manipulate cameras or switch between them.

(6) Listeners and speakers should appear to look one another in the eye.

The aim is to make both the audio and the video components of teleconferencing systems maximally useful for first-time users, without the need for special training or familiarity with the system.

This view rejects the familiar interactive voice systems that do not attempt to be realistic. The venerable "squawk box" systems that are familiar in industrial plants, railroad operation, or air traffic control apparently are very useful, yet obviously had no further design goal than that voice should be understandable. "Do you read me?" is all the user wants to know. Often even this goal is not really met, except for persons who use the equipment daily and are thus specially experienced in understanding the words that such low-quality loudspeakers scrape out. Visitors are often mystified as to how anything can be understood.

Beyond Face-to-Face Communication

There are also some good arguments *against* the closest possible simulation of face-to-face communication. A strong point is that many of our communication media cannot possibly simulate the face-to-face situation, yet have become very important to the development of our modern world.

Print-based or alphameric media cannot simulate face-to-face communication because they do not make possible the transmission of any of the face-to-face means. One cannot vocalize, gesture, use facial expressions or bodily position, dramatize (in the sense of acting out), sing, make music, or touch hands across the table, to name a few of the limitations.

It takes years of training before the medium of the printed page becomes transparent to the reader. Yet even fairly young children can "lose themselves" in a book, and for long periods of time — longer certainly than most TV producers would trust their programs to hold the same children's attention. Writing, publishing, and reading have elevated the words of face-to-face speech to new levels of expressiveness, and of course have extended them across time and space. Although there is nothing natural about either writing or reading, and a first-time user of the medium of print would find it unintelligible, the rewards of literacy are so great, and modern society so dependent on it at all levels of work, that the skills of reading and writing are considered mandatory before a person can become educated in any part of the world.

If years of study can be justified in learning literacy, why not also for other media? In our era of high technology, there are new skills that everyone is beginning to learn, to one degree or another. Nearly everyone can use a simple camera or a tape recorder. Many can use typewriters to some extent at least. If the need is great enough and the motivation is high, learning the operation of some pretty high technology is possible. After all, any teenager is an avid learner when it comes to the skill of driving a car. In the near future most young people will be reasonably familiar with the use of computers.

There are some communication skills, not too new, that are still not very widely learned. Even teachers who consider educational films valuable in the classroom often fail to master the skill of setting up a folding screen, let alone operating a projector, and unless a specially trained projectionist is available (often a pupil in the class,) this medium may be inaccessible. Television was more practical in this respect — all one had to do was to turn it on at the right time. Even so, the maintenance technician is heavily depended on. In many school districts where there is a television set in every classroom the most common cause of technical trouble turns out to be the accidental disconnection of power cords from their wall sockets.

Manufacturers have found that by making devices more and more automatic, with fewer controls for the user to learn, they can extend their use (and increase profits). Cameras are setting their own exposure and, with simple sonar equipment, can even detect the distance to the subject, set the focus accordingly, and turn on a light in the view finder when it is all right to press the button. The tape cassette, first audio and then video, made the tape recorder a commonplace in the home. Film projector manufacturers competed with videotape in schools and industry by developing cassette film projectors.

Operating skills are becoming less and less needed in the use of communication technology. Computer scientists have already devised machines to read print, although handwriting is still a bit elusive. They are working hard on the problem of designing computers that will understand human speech. Computers can talk already, by selecting recorded vocal sounds and putting them together in the necessary sequence to make words. But understanding what is said to them is different. This is a far more difficult assignment than any other computer development to date; if successful, it will eventually make it possible to bypass the skills of reading and writing. The technology is already here, for example, to construct a combination video camera and computer that can look at a road sign, for instance, and tell you what it says — out loud. If the sign is in Polish, the computer will tell you the equivalent in German, or English. Reading and writing can be accomplished today by machine; it is the understanding of speech that is difficult. So it will take a little longer.

It will be possible eventually to combine speech and writing in new ways. You will be able to speak to a computer that will turn your words into digital alphameric form, index your statements, print them out, and transmit them to other terminals according to your oral instructions. And if the computer can do all these things with one person's voice, why not with a group of users in a conference, or in a teleconference? We may supplement or largely replace computer conferencing in which entries are typed, since it will be easier to make oral entries. Means will then be devised for oral editing. Keyboards could become little used. With the incorporation of computer graphics, printed support materials, and chalkboard drawings, oral computer conferencing will become audiovisual.

In late 1979, a Japanese company, Sanyo, announced the development of a "voice-controlled TV receiver," the first incorporation of a voice-recognition device in a consumer item. This innovation was capable of memorizing and recognizing 30 different words, so long as these were always in the same person's voice. While this technology is in a very primitive state and only simple applications can be made of it so far, 30 different commands without the use of a keyboard is a far cry from the 5 buttons of QUBE (see Chapter 5). If 10 of these spoken words are the names of the numerical digits, even this primitive device can open up new realms of communication. Whereas voice recognition may not for some time be able to simulate face-to-face interaction, it could find application in the control and operating aspect of communication systems. For instance, 30 spoken words could reduce by 30 buttons the complex computer keyboards that now contain as many as 72 buttons. This would reduce the complex computer interface to no more than the standard 42 letter, numeral, and punctuation buttons of the manual typewriter. In regard to audio and audiovisual communication, a simple voice-recognition system would allow a person to carry on interactive communication, controlling and operating it by voice commands, while the individual's psychomotor system is busy with some other activity, such as driving a car or flying a plane.

The Factor of Apparent Distance

In designing a video conference room, the monitor showing the distant conference group is generally placed at what is considered a comfortable distance, rarely any closer than a live person would be placed, and usually considerably more distant than this.

Most video conference systems are designed with the entire conference group close together so they can be covered by the same camera(s) and all watch the same monitor(s). This tends to make the viewing distance rather great. Notable exceptions to this are the Bell Labs teleconference studios in Murray Hill and Holmdel, New Jersey, and the AT&T Picturephone first-generation Meeting Service studios that once operated in Washington, New York, and Chicago (see Chapter 1).

The size of the viewing monitor and its distance from the people who view it may have a more powerful effect on their communicative behavior, and thus on the success of a teleconference system, than has generally been appreciated. The reason for this is that there is a strong relationship between the amount of space separating persons and the way in which they relate to each other. Personal conversations, for example, are not comfortably carried on at a distance, and formal presentations, appropriate to an auditorium, seem out of place between people in a smaller group at close quarters.

In analyzing the effectiveness of pictures, it is recognized that simple size is one of the most important factors. This understanding gave rise to large-format magazines and magazines with foldout pages, sometimes three and four times the size of a regular page. A pulchritudinous centerfold unfolded at reading distance gave the reader the vicarious experience of examining the woman as though she were only six feet away. A full-head close-up, in a magazine the size of *Life,* can be larger than life-size. A life-size reproduction creates a virtual distance that is exactly the same as the actual viewing distance. In the case of printed material (book, magazine, or newspaper), the life-size photograph puts the viewer virtually within a foot or two of the subject, something that can feel uncomfortable if the pose is a formal one or the person is someone you might not want to come that close to.

Proxemics

Edward T. Hall, in his 1966 book *The Hidden Dimension,* introduced the term "proxemics" to cover the study of the use of space by human beings. The distance that separates people when they communicate has a marked effect on their behavior. As a first approximation, open to improvement as later investigations shed more light on the subject, Hall identified four distance zones — intimate, personal, social, and public — dividing each into a near and a far phase (see Table 3). He also observed that these zones, and the behavior associated with them, differ between cultures, even between different European countries.

Except for the first few phases, it is primarily visual information that is altered as people move from one zone to another. They can see each other better, and hence understand each other's facial and bodily expressions better, at close distances. When the range becomes less than 1.25 meters (4 feet), which includes the personal and intimate zones, other senses become important: Body odors, perfumes, the sounds of breath, heat from the other's body, and so forth begin to play a part in communication, albeit subconsciously in many cases.

Conferees who sit across a table from each other (when, for example, they come from different organizations) are accustomed to communicating in the zone of social distance. When they sit side by side (as members of the same team will do) they are in personal distance. Rarely is a conference table so large that anyone finds him- or herself at public distance (farther than 3.75 meters) from any of the other participants. Hall described social distance as the zone of normal voice level. When at a personal distance or closer,

TABLE 3 Hall's Zones of Communication

Zones	*Estimates*	*Metric Approximations**	*Characteristic Behavior (of Americans)*
Intimate distance			
near phase	touching to 6 inches	touching to 15 centimeters	Conversation tends to be in whispers.
far phase	6 inches to 18 inches	15 centimeters to .5 meters	
Personal distance			
near phase	18 inches to 2.5 feet	.5 meters to .75 meters	Reserved for people with whom one has close relationships. Can comfortably discuss subjects of a personal nature.
far phase	2.5 feet to 4 feet	.75 meters to 1.25 meters	
Social distance			
near phase	4 feet to 7 feet	1.25 meters to 2.25 meters	For discussing impersonal subjects of a business nature. Relations tend to be more formal. Voice levels rise.
far phase	7 feet to 12 feet	2.25 meters to 3.75 meters	
Public distance			
near phase	12 feet to 25 feet	3.75 meters to 7.5 meters	Tendency to project the voice. Formal speech-making. Projection of personality by good communicators.
far phase	25 feet to beyond	7.5 meters to beyond	

*In most cases these approximations are to the nearest .25 meters. This makes all conversions within about 6 percent, except for 18 inches, which actually converts to .457 meters. The round figure, .5 meters, is 9.3 percent larger than this. The figure is not used in the accompanying discussion, in any case.

one tends to lower one's voice; at public distance one tends to talk louder.

Proxemics and the Visual Media

To the question "How far distant, in effect, are the people you see on the monitor?" some participants might give their estimate of the distance between themselves and the TV screen. Others might think a little further and add the distance at the other end of the line between the subject and the pickup camera. These are not the only factors in apparent proximity, however. We are all familiar with telephoto lenses that make a distant subject seem close, and we know that television screens come in a wide range of sizes.

There are at least four factors that determine the proximity of people seen on a TV screen: (1) distance of the pickup camera from the subject, (2) focal length of the camera lens, (3) size of the display screen, and (4) distance between the display screen and the viewer. The first two factors determine the compositional content of the shot and can be combined for our purposes into a single factor, which we may call "shot content." the third and fourth factors, screen size and viewing distance, can be combined into a single very convenient measurement: distance from the screen in screen widths.[1]

These four factors, reduced to two by the combining we have just done, determine the size on the retina of the image of a person's face resulting from looking at the face on a TV screen. Logically, a larger retinal image would be perceived as a person in closer proximity; a smaller image as a person farther away.

The face on the TV screen subtends an angle at the eye (see Figure 23) that is the same angle that the face of the real person would subtend if he or she were there in the flesh but somewhat farther away. How much farther away would depend on the size of the image on the monitor and the distance it is from the viewer. This is what we will call the "virtual distance." It will be greater than the viewing distance to the television set unless the image on the TV monitor is life-size or larger. If it is over life-size, virtual distance will be shorter than viewing distance. If the image on the monitor is exactly life-size, then the actual viewing distance between viewer and monitor will be the same as the virtual distance of the image.

By our definition, virtual distance is determined solely by the size of the retinal image. There seem to be several other factors that affect *apparent* distance, however. The image of a face lying on a television screen is a two-dimensional image and is comprehended as such. It is a picture, surrounded by and contained within a frame, and we have special ways of looking at pictures. We project ourselves, so to speak, into the picture, concentrating all

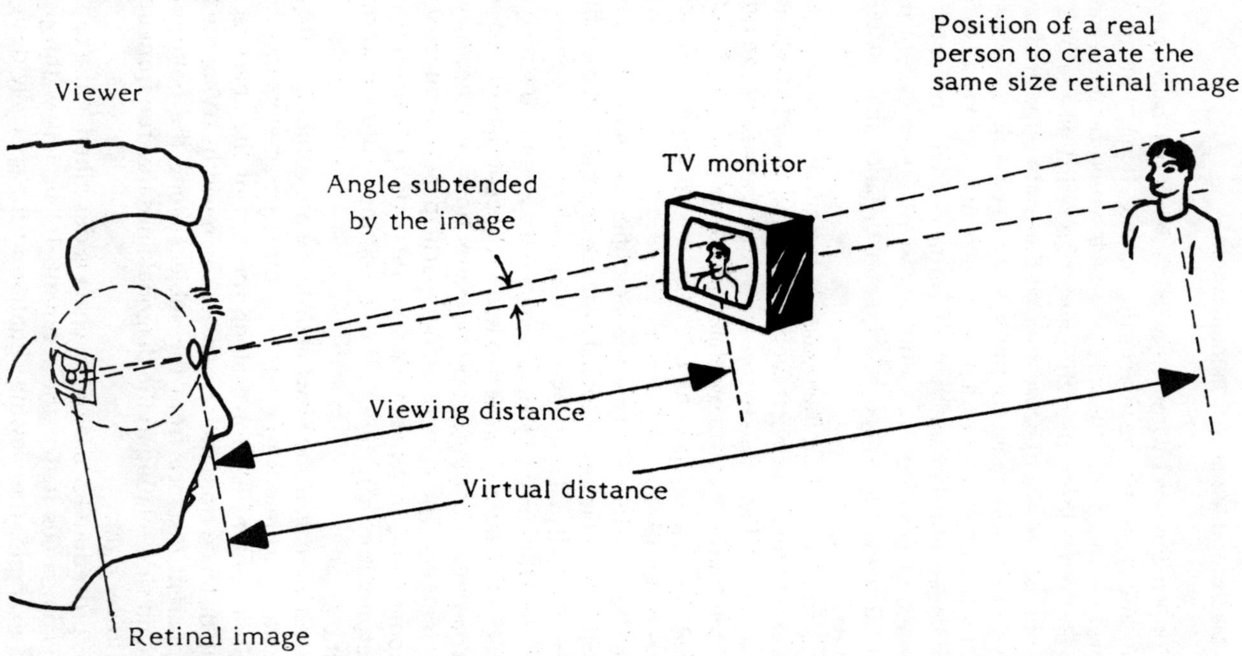

Figure 23: When the TV image is less than life-size, its virtual distance is greater than viewing distance.

attention on the area within the frame while virtually ignoring surrounding areas and the images they create. It might be guessed that a retinal image of a picture of a face would be perceived as larger, and thus closer, than the same size retinal image of a real face. There is evidence, however, that the exact opposite may be the case. In an experiment conducted by Barry Stapley (1972), it was consistently found that the television image of a person had to form a much larger retinal image than the actual person in order to have the same apparent proximity. In other words, the *apparent* proximity of a television image is a greater distance than its *virtual* proximity.[2]

It is also possible that the quality of the audio originating from the same subject can affect its apparent distance, and there can be other factors. All we can say for sure is that the size of the retinal image determines its visible detail and visible detail carries information that certainly acts as one stimulus to appropriate communication behavior.

The chart in Figure 24 (and the simplified version in Figure 25) relates the three factors of virtual distance and makes it possible to determine the virtual proximity of a respondent when seen on a TV screen. We will not say "virtual distance between two interacting persons," because an interactive telelinkage consists of two complete channels operating in opposite directions, and they may not be symmetrical. The camera on one participant may be closer, or use a longer lens, than the other, and one participant may be using a larger monitor than the other or be sitting a different distance from it.

Since screen size and viewing distance are combined into one factor, the chart needs to relate only three parameters instead of four. Shot content is shown along the top of the chart, actual viewing distance from the screen in screen widths (W) at the left, and apparent distance from the subject along the bottom and right-hand side relating to the diagonal lines that run from upper left to lower right.

The procedure for use of the chart is described in the caption that accompanies it. For the moment, let us follow through a typical example, the viewing situation in the specially designed Confravision studio.

The Confravision Example

In this system there are two shots available, a long shot showing five people in a horizontal line, and a medium shot showing three. The Confravision system is symmetrical, so what is determined for the London studio will hold true also for Glasgow, Manchester, Birmingham, or Bristol, whichever two happen to be in communication, assuming that both are using the same shot.

Starting with the Confravision long shot, a row of 5 people with heads 10 percent of screen height, we follow down the 10 percent column and stop at the appropriate viewing distance, 5 W. (Confravision users sit 9 feet away from a 24-inch monitor, which is actually 22 inches wide; 24 is the diagonal measure.) Laying a straightedge on the chart so it passes through zero and the intersection of 5 W and 10 percent, we find that it intersects the scale for virtual distance, along the bottom, at about 60 feet, well into far public distance.

The closer shot in Confravision contains 3 persons and is indicated most closely by the second drawing from the left at the top of the chart. Following down the 20 percent column to the 5 W line and laying the straightedge between that point and zero, we find we are now virtually as close as 25 feet, mid public distance.

If we believe that confronting our distant participants at a virtual distance of no more than 7 feet would increase the effectiveness of video conferencing, how could we redesign the Confravision system to meet this criterion? We have three choices, each of which would be more expensive that what was originally done. We could: (1) use closer shots, (2) decrease the actual viewing distance, or (3) use a larger screen.

Let's examine the first possibility. How large should the faces of the televised participants be if they are to seem 7 feet away? Using tha chart this time from the bottom up, we start with the virtual distance of 7 feet, shown on the right-hand side of the chart. We then follow the diagonal up until it crosses the 5 W horizontal line, and then up the column to find what camera shot will be needed. It is a full-head close-up, 70 percent the height of the screen. To cover a 5-person conference in shots like this with fewer than 5 cameras would require a camera operator, either on the camera or remotely,

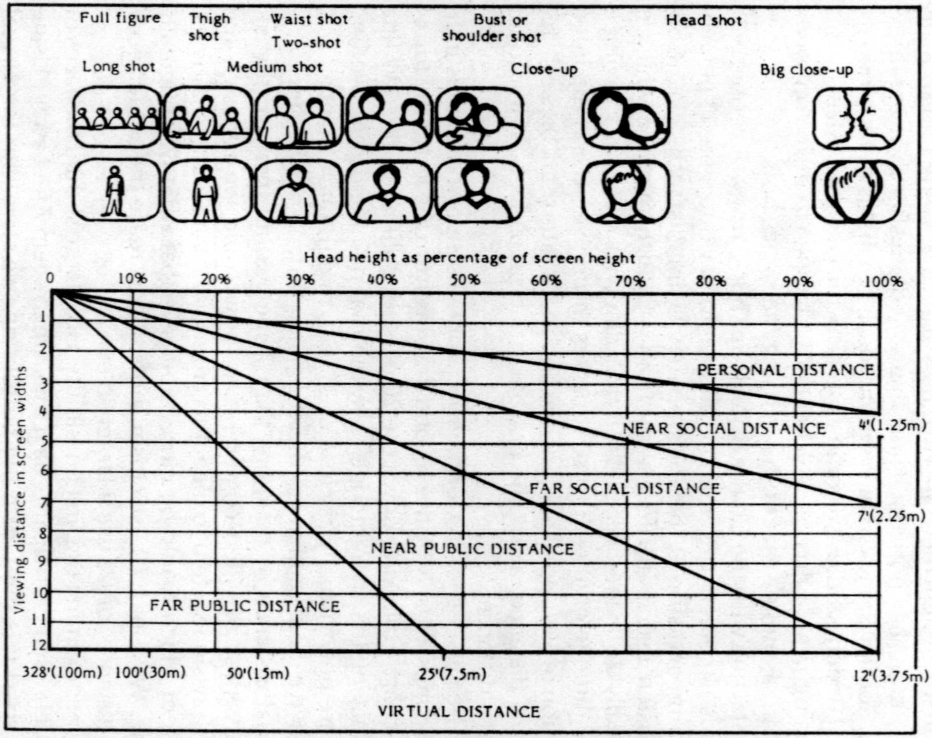

Figure 24: Shot content, viewing distance (in screen widths), and virtual distance related.

NOTE: How to use this chart—

(1) To determine virtual distance if viewing distance and shot content are given:
  (a) Find shot content (top of chart). Some typical examples of shot composition are sketched. If none of these is representative, follow the scale of head size just below, estimating the size of a person's head in relation to the full height of the frame.
  (b) Locate intersection of column representing shot content with horizontal line representing viewing distance.
  (c) Follow diagonal lines down and right; read virtual distance at bottom or right-hand side of chart.

(2) To determine viewing distance if shot content and virtual distance are given:
  (a) Find shot content.
  (b) Find virtual distance at bottom or right-hand side of chart.
  (c) Locate intersection of shot content column and diagonal representing virtual distance (or range thereof).
  (d) Read viewing distance (or range thereof) in screen widths at left.

(3) To determine shot content if viewing distance and virtual distance are given:
  (a) Find viewing distance (or range thereof) at left of chart.
  (b) Find virtual distance at bottom or right-hand side of chart.
  (c) Locate intersection of viewing distance line and virtual distance diagonal.
  (d) Read shot content at head of column.

EXAMPLE 1: My viewers are to be seated between 4 and 8 screen widths from the screen. I intend to use 2-shots. What will the range of virtual distance be? *Answer:* The range of near public distance—from 12 feet for the closest viewers to 25 feet for the farthest.
EXAMPLE 2: I want to simulate the far phase of social distance using tight 3-shots (3 people close together in the frame). How close to the screen must my viewers sit? *Answer:* 2 to 3 screen widths.
EXAMPLE 3: My viewers will be 6 to 12 screen widths from the screen and I want to simulate near social distance. What shots must I use? *Answer:* Extreme close-ups. Only the closest viewers will experience near social distance even when faces are full screen height.

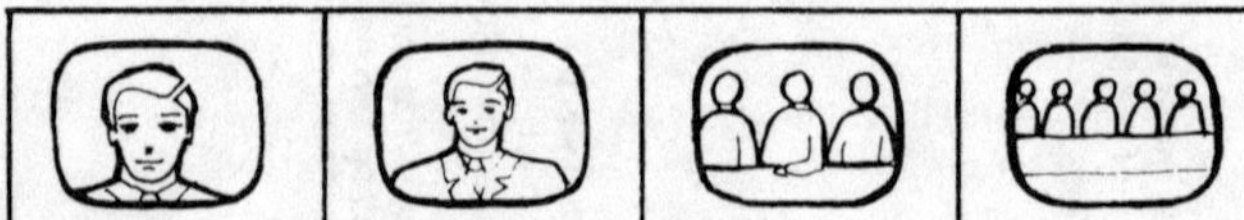

Viewing distance in screen widths.		Head shot	Shoulder shot	3-shot	Long shot
	Head height*	80%	50%	25%	12%
	2W	Mid personal distance	Far personal distance	Far social distance	Near public distance
	3W	Far personal distance	Near social distance	Far social distance	Mid public distance
	4W	Near social distance	Far social distance	Near public distance	Far public distance
	5W	Near social distance	Far social distance	Near public distance	Far public distance
	6W	Mid social distance	Far social distance	Mid public distance	Far public distance
	7W	Far social distance	Near public distance	Far public distance	Far public distance

*As percentage of screen height.

Figure 25: Virtual distance for various combinations of shot content and viewing distance.

as in the case of the MRC-TV system (see Chapter 1). In TV studio production, 2 cameras with operators are generally used on discussion programs, often 3.

The other alternatives, a shorter actual viewing distance or a larger viewing screen, both decrease the viewing distance when it is measured in screen widths. Taking actual viewing distance first, let's use the chart to determine how close to the screen we must be so that the people in the 5-person shot are virtually 7 feet away. Starting at the 5-person shot we go down the column until we encounter the diagonal representing 7 feet virtual distance. We would have to be less than 1 screen width away. In the case of the 3-shot, represented by the next column, it would be 1.5 screen widths. This would mean that each Confravision user would need a separate 24-inch monitor within 2 or 3 feet of his or her eyes.

The last possibility is to enlarge the viewing screen. What size should the viewing screen be for people in the Confravision 5-shot to seem 7 feet away? If the viewing distance is 1 screen width, and the viewers are 9 feet away, then the screen must be 9 feet wide. This is for the 5-shot, the mode in which Confravision was found to be used about 60 percent of the time. For the 3-shot the viewing distance would be 1.5 screen widths; a virtual distance of 7 feet could be obtained using a screen less than 5 feet across.

These examples are given mainly to demonstrate the use of the chart. Of course, more than one of these changes could be made at once, so none would have to be extreme. If medium shots were used, for example, where head heights were 30 or 40 percent of screen height, and participants sat within 3 screen widths, virtual distance could be between 7 and 10 feet. When viewing distance becomes too close, however, direct eye contact may suffer. We will come to this problem shortly.

The Banker's Trust Example

Confravision is not, however, the worst example. In an earlier chapter some bank systems were described in which conference rooms in downtown New York were joined by TV to similar conference rooms in the same banks' midtown offices. The conference room seen by one of the authors during the late 1960s was at the Bankers Trust Bank in midtown Manhattan. To the best of his recollection it sat eleven people, five on each side of a long, slightly oval table, with the chairperson at the end. At the other end of the room, possibly a meter and a half from the end of the table, a large TV monitor was built in, flush with the wall, with a camera

NOTE: This image is reproduced here in the correct size so that when held at a standard reading distance (38 cm or 15 inches), it creates the same size retinal image as the bank's monitor did in the eyes of the meeting chairman.

Note the distance you are holding this page from your eyes. This is your reading distance. Normally it would be around 38 cm, but people vary. If your reading distance is within 5 or 10 cm of this figure, you can still perform this experiment. Maintaining this reading distance, raise the book until it is level with your eyes. The drawing has been placed near the edge of the page to make the next step easier. Holding the book at the right height and distance, turn until you are facing out a window or down a large room or corridor. Look alternately at the picture and at the real scene beyond it. It may help to close one eye. The drawing should look about the same size as a large TV set about 4 meters away. Check again to make sure you are holding the page about reading distance from your eye. If you can pick a spot about 4 meters distant and visualize a TV set there with this drawing on it, you will have an idea of how small the facial images of the bankers must have been. Continue to hold the drawing at reading distance.

Now forget the drawing as a whole and consider only the image of the chairman at the head of the table. Have someone walk away from you until his or her head is the same apparent size as that of the chairman in the drawing. If you don't have a confederate, perhaps there are other people in the scene who will help you estimate where a person would have to stand to be the same apparent size as this drawing of the chairman. Now you have matched in size your retinal image of the drawing of the chairman and your retinal image of the live and distant person. How far away is the real person? This was the virtual distance of one chairman from the other in the Banker's Trust teleconference system. According to Stapley's (1972) experiment, the *apparent* distance would have been twice as great as this.

Figure 26: Banker's Trust teleconference image.

port just above it. (There was a second camera mounted in the ceiling over the table for displaying graphic materials.) The main camera provided only one shot, a wide angle showing all eleven persons at the table (see Figure 26).

Viewing distance was different, of course, for each pair of conferees seated across from each other down the table; the chairman at the end of the table had the most distant position. (He was also, probably, the highest ranking member of the group, whose evaluation of the system may have counted for more than that of all the others put together.) The chairman was perhaps 13 feet (4 meters) from the monitor. On the monitor was a similar image to that shown in Figure 26, showing another eleven people or so in a similar conference room. The face of the other chairman at the far end of his table must have been less than one-tenth the height of the frame. With the chairman at 7 screen widths from the monitor, when the face of the other chairman was about 4 centimeters high, the virtual distance between them must have been at least 80 feet (24 meters). The nearest participants, 3 screen widths from the monitor, may have been barely in virtual social distance from the nearest participant on the screen, but all the rest were at virtual public distance from all transmitted participants (see Figure 27).

After some eight or ten years of operation of this bank system, during which time the video equipment had become progressively less effective and less reliable, it gave out altogether and the system was continued as an audio-only system; the video was not considered important enough to replace. Small wonder — it must hae been contributing very little if anything to the conferences for which it was used.

We have said that there probably are other factors than virtual distance operating to create the illusion we are calling "apparent distance." For example, it is common knowledge that an exchange of glances (eye contact) across a room, exchanges of smiles, tilts of the head, and other non-verbal communication can reduce the effective distance between people in face-to-face interaction, and probably in video interaction as well.

There is also something to be considered about the role of audio in establishing an apparent proximity. If a person is visually in public distance but audially in personal distance, due to using a very close microphone, plus perhaps being received on a close speaker or earphones, what will be the net result?

Edward T. Hall (1966: 123) in his description of "social distance — far phase" suggests that "raising the voice or shouting can have the effect of reducing social distance to personal distance," but he gives no evidence or argument to support this opinion. There is,

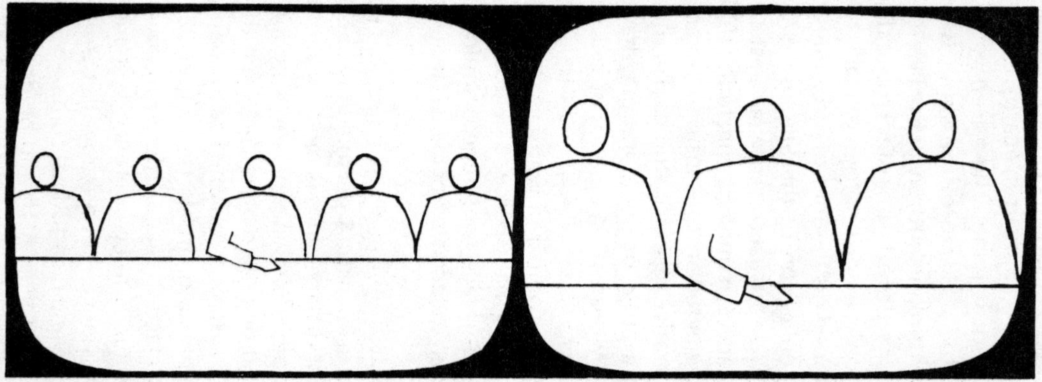

NOTE: When this page is held 15 inches from the eye, the retinal images produced will be the same size as those experienced by the users of the systems represented. The top picture represents the typical Picturephone image as seen from a distance of 3 screen widths; the bottom pictures represent typical Confravision images as seen from a distance of 6 screen widths.

Figure 27: Comparison of the virtual size of the facial image as seen on the original Picturephone unit and as seen on British Confravision.

however, a major difference between a raised voice or shout coming from a distance, and a normal voice coming from nearby, even if they both arrive at the same level of volume. The raised voice, moreover, is associated with far social distance, the shout with public distance. Hall was of course thinking entirely of the face-to-face situation; he had not considered the electronically reproduced voice (nor the electronically or photographically reproduced picture) when he developed his theories of proxemics.

Whether or not there is any replication of the Stapley (1972) finding that apparent distance is twice virtual distance (at least 2.5 m, near the midpoint of social distance), it must be concluded that proxemics has been seriously neglected in the design of many teleconference systems and comparative research studies. This has not been a minor oversight, but a gross neglect, and may have been partly responsible for the slowness of acceptance or lack of use of many teleconference systems.

If video teleconferencing is to be compared with face-to-face interaction, it seems mandatory that as many perceptual factors as possible should be the same. A face-to-face conference should be designed that simulates the video teleconferencing system to which it is to be compared.

In the 1973 CSG study of Confravision use, it was found that 60 percent of the time the system was used in the 5-person (long-shot) mode. We have seen that apparent distance in this mode is about 20 meters. Find an empty ballroom somewhere, seat 5 people behind a table at each end with 60 feet (20 meters) between the tables, and then hook up an audio system so no one has to shout, and the situation should be roughly comparable (see Figure 28). What advantages are there *now* to face-to-face conferencing over teleconferencing? To video over audio teleconference systems? Unfortunately, this comparison has never been made. If it is impractical, or illogical, to match video and face-to-face conferencing by increasing face-to-face distance, at least it should be done by increasing the size of the video image. In the following section some conference systems are proposed (see Figures 31 and 32) that put communicants virtually within 2.5 meters of one another, and thus might be used to simulate a face-to-face meeting in which people are within social distance.

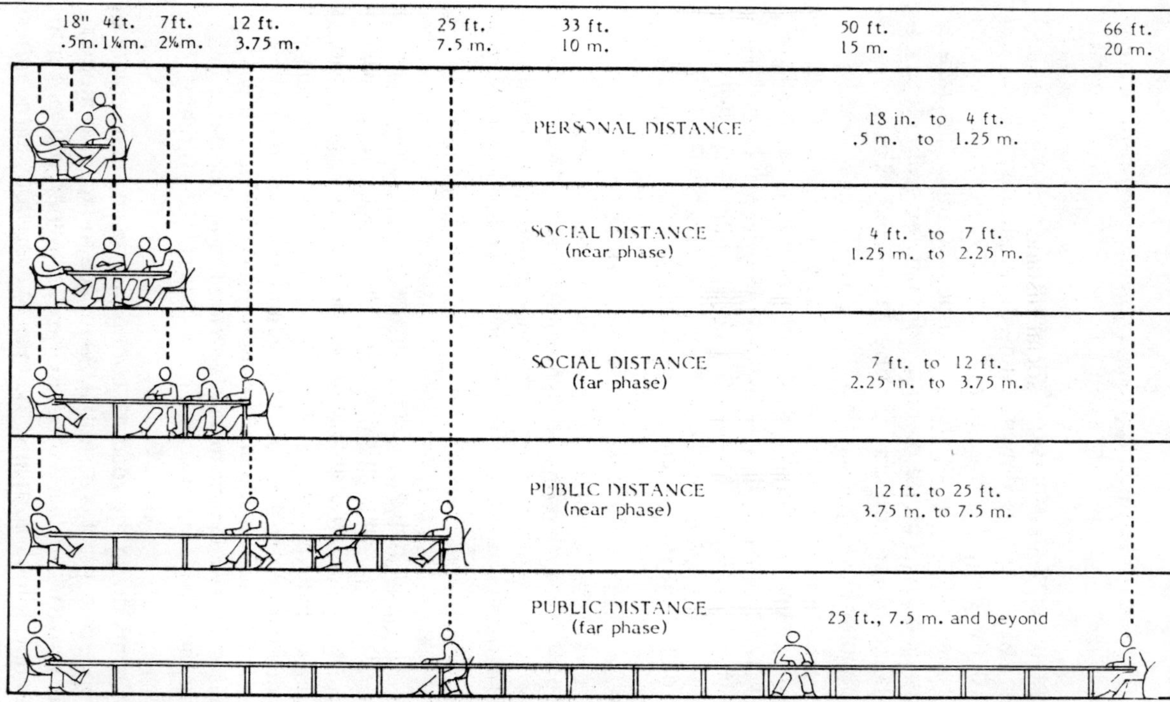

NOTE: The two examples at the bottom simulate typical virtual distances in video conference systems. The Misar II system most closely resembles the middle drawing.

Figure 28: Face-to-face conferences at various actual distances.

The Factor of Eye Contact

Direct eye contact is no problem when you make a picture of someone, whether it is drawn, painted, or photographed. You simply ask the subject to look directly at you, or at the camera, as the case may be, and he or she will appear to meet the gaze of everyone who looks at the picture, from whatever direction it is viewed. One of the authors describes his experiences with this phenomenon.

> I can remember as a child of four or five observing this effect for the first time. A magazine featured a picture of a man that I did not like. He was glaring at the camera and I did not like to be looked at in this fashion. So I did what I would do in the face-to-face situation — I moved off to the side, or turned the magazine, so that I would no longer be in his line of vision. Alas, he was still glaring directly at me. Even when I was looking at him from such an extreme angle that he was squeezed down to be hardly visible, his eyes where still fixed malevolently upon me. That's funny, I thought, as I turned the magazine face down. I accepted the phenomenon as part of the world of nature, and forgot there was anything odd about it.
>
> Thirty years later I was training teachers to use educational television effectively. "Look directly into the camera," I said, "and you will be making direct eye contact with everyone in your TV audience 100 percent of the time. In the classroom, where you have a class of thirty and you want to make eye contact, you have to keep looking around; you can make eye contact with any one student only one-thirtieth of the time. How important *is* eye contact? It keeps your class with you, it gives you constant feedback, it personalizes your relationship to the class. Now I am introducting you to a medium that can increase your eye contact by thirty times!"

Sixty years down the road it's coming up again; in interactive television, eye contact is essential. As in face-to-face communication, we want to "look each other in the eye." When you look into the camera you look right into the eyes of everybody at the other end — everyone at once, not one at a time as you must do in a face-to-face meeting.

But what if you don't want eye contact with everyone at once? What if you want to exchange glances with just one person, perhaps while someone else is talking — a private exchange of smiles or knowing looks. It cannot be done. Via TV you look at everybody or nobody. There is nothing in between. It's like audio communication in a face-to-face meeting. Unless you whisper or greatly lower your voice, your words are heard by all. The only way to exchange private glances in today's teleconferencing would be to have a separate private TV system, with separate cameras, transmission channels, and monitors for each two people who may want to interact.

Why is eye contact all or nothing? It's because the television image is two dimensional. It gives us only a *simulation* of depth. When a person looks at you from a TV screen, his or her line of vision, if it existed, would come forward in the depth dimension. But since the picture has no depth dimension, there can be no line of vision.

When you look at a three-dimensional hologram picture the effect is different. Move a little to the side and you begin to see the subject from a side angle. You can get out of the line of vision of a holographic image of a person as easily as you can from that of a real person. It's the same as getting out of the line of vision of a statue. You can stretch a string directly along the line of vision of a person in a hologram. With a two-dimensional image there is no such dimension in which to run your string. Your string would have to go everywhere — to everyplace from which the image can be seen.

Someday in the future a large-screen three-dimensional image will go most of the way toward a realistic simulation of face-to-face relationships. In such a future telemeeting people could direct their gazes where they wanted, and nowhere else. A man at the far right of one conference table could exchange smiles with a woman at the far left of the distant meeting group and no one else. He could direct a statement or a question to a specific member of the other group merely by looking at that person. Three-dimensional sound could be added to the picture so that voices would emanate from appropriate parts of the screen.

NOTE: Photo courtesy of Misar Industries.

Figure 29: Toward the telehandshake.

But where is the camera that picks up the outgoing picture of the local participants? This is a hypothetical possibility to begin with, so the question can be glibly answered. With holographic systems, cameras as we think of them aren't used anyway. We are assuming a laser pickup of the images at each end of the line; the screen will carry a laser display. Our big assumption is that laser pickup and laser display can operate in the same space. Until that feat has been demonstrated, direct eye contact by a televised participant who is facing a "large screen" (meaning he or she is less than four screen widths distant) will continue to be a problem.

Today, however, we do have the large screen in *two* dimensions, which is an intermediate step that Misar Industries has already taken, placing their people, in support of this theory, just over four screen widths from the screen. Figure 29, the attempted handshake, was set up and photographed by Misar for use in promoting the sale of their teleconferencing system. In a sense, however, this picture is more representative of a future three-dimensional system than of today's best teleconferencing configuration. Let us explain why.

Looking at this picture, it appears as though the two participants are meeting each other's gaze, even though they can't quite do the same with the handshake. Actually, however, the near participant does not see his distant counterpart looking him in the eye at all. The man in the background is a two-dimensional image on a screen, hence he appears the same from all angles. The near participant would see exactly the same image that we see, a man looking off to the left. The shot was specially staged to create this dramatic effect; during actual use of the system such a shot could not be taken. The Misar II system, like any other video conferencing system, has the distant participant looking almost directly into the camera and hence at everyone at once, including any photographer who might take such a picture.

There is another point to be made about this effective picture, concerning the proximity of the two people. The image is produced, in this system, on a fifty-inch front-projection screen (this is a diagonal measurement; the screen size is thirty by forty inches), by a projector located under the conference table. The near participant appears to be about six feet from the screen, which amounts to less than two screen widths. The screen image appears to be about life-size. With a life-size image, virtual distance and viewing distance are the same, so the two participants appear to be six feet apart, which is "near social" distance — not close enough to actually connect in a handshake, but the normal distance across a conference table. What is shown is almost an ideal teleconference image, except that the camera, which can be seen just above the screen, is displaced, at this close a viewing distance, at a vertical angle that may be so great that participants may appear to be looking down. In the Misar II system (see Figure 15, in Chapter 1), the screen is placed about four screen widths from the participants to keep the parallactic angle between screen and camera as small as possible. In staging this shot, the photographer has had the near participant step around in front of the table, which can be seen at his back, across the bottom of the photo.

Eye contact is intimately related to the camera-screen problem. Because, in interactive TV, a person's image must be picked up at the same time he or she is viewing the image of the distant communicant, there must be both a viewing screen and a camera before each participant. Wherever the camera is placed, the intercommunicator will ignore it, so powerful is the compulsion to look at the person to whom one is talking.

Knowing that everyone will look at the monitor, system designers try to place the camera as close as possible to the screen so that it will *seem* that people are looking into the camera. If sufficient closeness is not possible and a distant communicant does not seem to be looking you in the eye, of course all is not lost. You still can see the person you are talking to and watch his or her responses to what you say, and, when it is the other person's turn to speak, have the benefit of his or her gestures, facial expressions, and the like. But just as it is disconcerting to talk to someone face to face who is not looking at you, neither does it seem natural with interactive television. This is especially true when you are asking someone a question. People *will* use a system that is set up that way, and some may not be aware of what is missing, but any such system can be greatly improved if direct eye contact between speaker and listener can be achieved.

How close to the monitor should the camera be placed? The ideal, of course, is that both be on the same line of vision. Naturally, one cannot be in front of the other, so in most of the systems we have described, the camera is simply as close to the monitor as it can physically be placed. In some cases this is just above the monitor, in other cases just below or to the side.

In the British Confravision system the camera is above and considerably behind the monitor. In the Picturephone system in which three conference cameras are used, they are just above the monitor in some locations and just below in others. In the MRC-TV system each conference room camera was mounted above its monitor on the same portable stand. Because pan and tilt apparatus was needed below the camera for remote operation, the camera lens could not be brought closer than 20 inches from the center of the monitor. This could have been reduced to 16 inches had the camera been placed below the monitor so that the pan-and-tilt apparatus did not intervene.

In the Reading system cameras were mounted on tripods and dollies (three-legged stands with wheeled support), which were manually operated by a standing camera operator. There are no data immediately available on this, but it is a fair guess that the camera lens must have been at least three feet (one meter) from the center of the monitor, directly to the side. In the Los Angeles Picturephone Meeting Service room the three conference cameras were placed just above a 21-inch-wide monitor, making the lenses about 14 inches above monitor center (see Table 4 for comparison of these systems).

The actual distance is not the important factor here, however; instead, it is the parallactic angle subtended at the viewer's eye by the camera lens in one direction and the monitor in the other (see Figure 30). This angle depends on the distance of the participant from the monitor as well as on the separation between monitor and camera. The viewing distance at the other end of the line may contribute somewhat to the illusion of eye contact, in that a greater viewing distance makes it more difficult to see a person's eyes. A distant viewer may feel that a person on the screen is looking directly at him or her, while a person closer to the screen might be aware of a slight deflection of gaze.

The problem becomes greater if participants are placed closer to the monitor, or if large-screen displays are used. Let us say, for example, that the 21-inch monitor in the Picturephone meeting room is replaced by a large projection screen, twice the monitor width (a little over a yard wide). Instead of being 5 or 6 screen widths distant, the same conference positions would immediately be within 2.5 to 3 screen widths. This would halve the virtual distance of the distant participants, even if no changes were made in the pickup at the distant location. A head-and-shoulders shot, for example, on such a screen, would make the person appear to be in near social distance, where as before he or she would have been on the borderline of public distance. A shot wide enough to hold 3 people would still put them in far social distance, whereas before it would have put them into mid public distance (see Figure 25). From the standpoint of proxemics such a wide screen system would be highly desirable. But what about parallax?

We have doubled each dimension of the screen. No matter on which edge of the screen the camera is placed, the parallax is

TABLE 4 Comparison of Video Teleconference Systems in Factors Affecting Eye Contact

	British Confravision	*AT&T Picturephone Meeting Service*	*MRC-TV*	*Reading*	*Misar*
Number of conference cameras	1	3	1	1	2
Position in relation to incoming monitor	above	above	above	side	above
Monitor size					
diagonal[a]	24 (61)	23 (58)	23 (58)	23 (58)	50 (127)
width[a]	19 (49)	20 (46)	18 (46)	18 (46)	40 (102)
height[b]	14.4 (37)	13.8 (35)	13.8 (35)	13.8 (35)	30 (76)
Average viewing distance					
in feet (meters)[c]	9 (2.74)	9.5 (2.9)	14 (4.3)	10 (3)	14 (4.3)
in screen widths	5.7	5.7	9.3	6.7	4.2
Actual displacement between					
lens center and monitor center[a]	24 (61)	14 (36)	22 (56)	36 (91)	18 (46)
lens center and ⅔ monitor height[b]	21.6 (55)	11.7 (30)	19.7 (50)	n.a.[d]	13 (33)
Parallactic angle (between lens center and ⅔ monitor height)	11.3°	6°	6.7°	16.7°	4.4°

a. To the nearest inch (centimeter).
b. To the nearest .1 inch (centimeter).
c. To the nearest .1 foot (meter).
d. Side displacement.

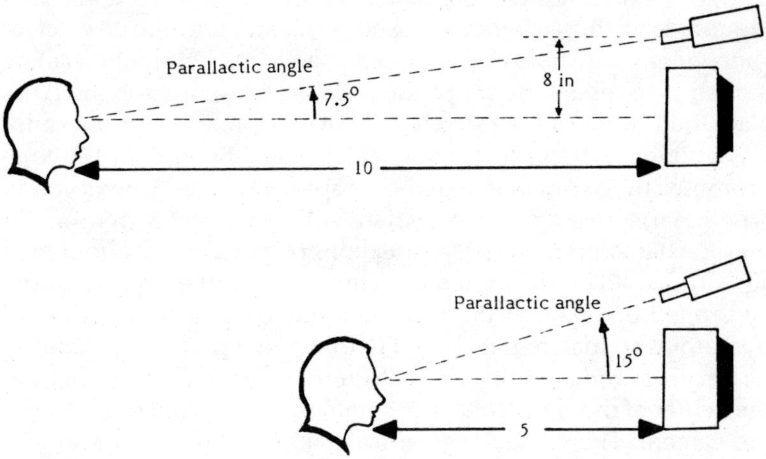

NOTE: When a speaker is 6 screen widths from the camera-monitor assembly, as shown in the top drawing, the parallactic angle between the two is 7.5 degrees. As the person fixes his gaze on the monitor, it will seem as though he is looking into the camera. If, as shown in the bottom drawing, he is half the distance from the monitor (3 W), the angle will be twice as great (15 degrees), and it will not seem to the viewer that the person is looking into the camera (unless the viewer is so far back from his monitor that the pupils of the speaker's eyes are not clearly visible).

Figure 30: Parallactic angle.

nearly doubled (nearly because we have doubled only the screen, not the space required for mask and cabinet or for the camera body and other structure, all of which also intervene between the camera lens and the center of the monitor). Assuming an 80 pecent increase in parallax, an angle that was 8.5 degrees now becomes over 15 degrees.

One-way television has long faced a problem very similar to this. When news and commercial announcers, for instance, face the cameras, they prefer to make eye contact with the audience as much as possible, only occasionally glancing down at their scripts. In order to do this they must either memorize what they have to say or have the script in legible form somewhere in the direction of the camera lens, preferably as close to the lens as possible so that they will seem to be looking directly at the audience the whole time.

After some years of holding cue cards to the side of the camera, or right next to the camera lens, prompters were designed that mounted on the camera and used a transparent mirror or clear sheet of glass. The glass was placed at an angle before the camera lens so as to reflect the script above it. The camera was behind the glass, but, because the reflected script was lighted, only the script was visible — the user did not see the camera lens. In the teleprompter the script was typed on paper about 18 inches wide in type so large that only 8 lines where visible at once, with room for only 25 characters (and spaces) per line. This meant that about 25 or 30 words could be visible at once. The script unrolled at the bottom and rolled up again at the top, remotely controlled by a special operator who matched its speed to that of the performer. Parallactic error was reduced practically to zero. When looking directly at the center of the script area, the commentator looked directly into the camera. The camera-viewer axis and the monitor-viewer axis were one; hence this kind of system can be called coaxial.

Of course there is no purpose to be served in interactive TV by reading prepared script or reciting memorized lines so that viewers think you are talking spontaneously. But the script in such a device could be replaced by a monitor showing the distant participants. If a system similar to that at Bell Labs (see Chapter 1) were used, with a separate camera and monitor for each pair of participants, a coaxial design would be practical; 2 people could be placed 3 screen widths from a 21-inch monitor, and the semitransparent mirror or plain sheet of glass required would be no larger than 27 to 32 inches square. It could be angled so it reflected a monitor below it rather than above in the fashion of the camera-mounted prompter. A separate camera-monitor assembly would be needed for each two participants. The camera, of course, would have little flexibility, even of zooming (see Figure 31).[3]

If a single large-screen projection is used, as in the Picturephone system, with six participants and three cameras, three screen widths will be a greater actual distance, and the intervening mirror will have to be quite large, not only to cover the larger screen, but to extend far enough left and right to cover the side angles from which the outer members of the group will view the screen. In either case, the conference table is slightly curved so that the local group can see each other conveniently and interact

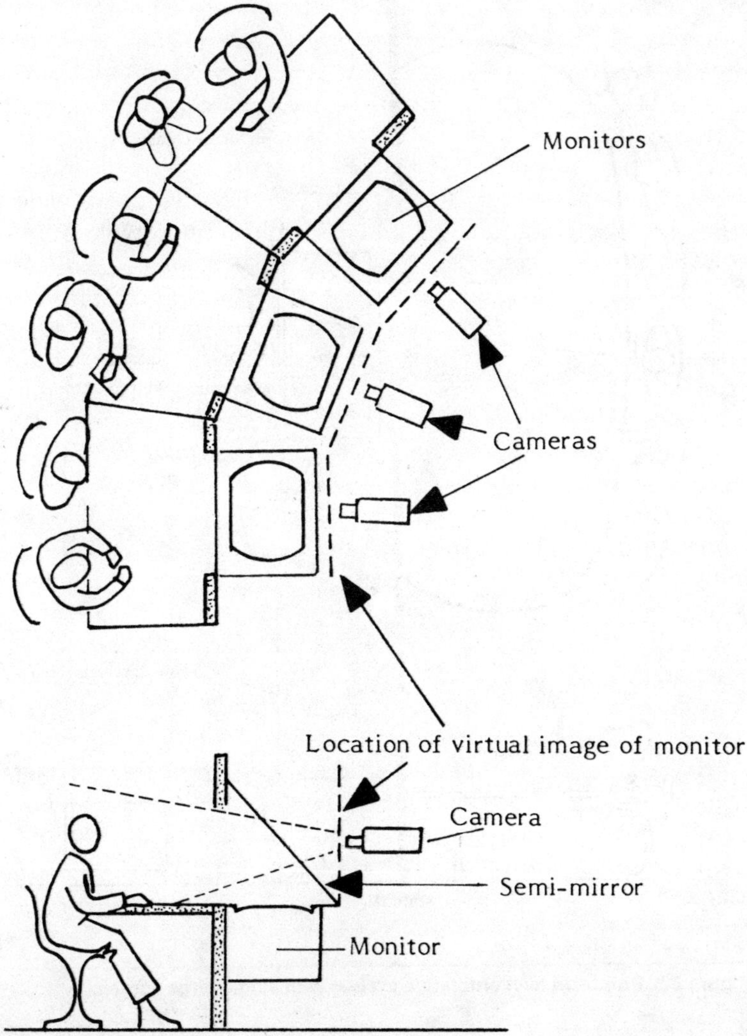

NOTE: Viewing distance = 3 W.

Figure 31: Proposed teleconference system with coaxial placement of cameras and monitors.

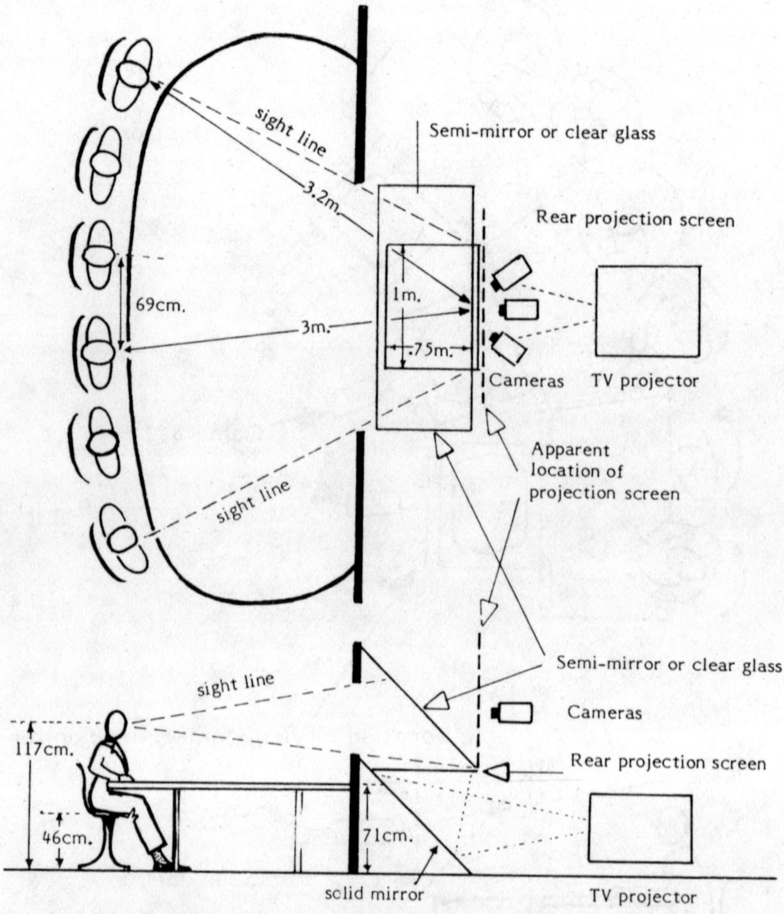

NOTE: Viewing distance = 3 W.

Figure 32: Proposed teleconference system with single large screen.

locally if they wish. The parallactic angle would be slightly greater in this single-screen system (Figure 32) for the people at the far ends only, since the cameras that would pick up their images would not be quite centered in the screen.

In either case, a viewing distance of three screen widths will make painfully evident such shortcomings, typical of the broadcast TV picture, as poor definition and color fringing, if a color system is used. Poor picture quality can interpose a media limitation between communicants using the system and reduce its transparency. It will have to be assumed, therefore, that the display used in such a system would be capable of picture quality well above the average of home TV reception, as it has been known in America at least, both in equipment design and in adjustment and maintenance.

Monitors

In the foregoing discussion we have considered only one monitor — the one that shows the incoming picture. This provides the essential visual contact with persons at distant terminals of the system who are alternately speaking and listening. In not all cases have both of these been possible. In the MRC-TV and the Reading systems, showing the person(s) listening to the speaker could only be accomplished by the use of a split screen. The importance to a speaker of seeing a listener has been stressed in previous pages.

Other pictures that a system user might like to see (particularly when speaking) are:

(1) the local outgoing image

(2) occasional glimpses of other terminals in a multiterminal system when they are in listening mode

(3) previews of additional picture sources at the local terminal, including:
   (a) a graphics camera
   (b) a slide pickup camera
   (c) a multipurpose camera
   (d) a videotape player

The monitor carrying the outgoing picture has in many systems apparently been considered of equal importance with the incoming picture. In the Confravision, Picturephone, and MRC-TV systems, both these monitors were equal in size and placed either side by side or one above the other. In the Reading system, on the other hand, no outgoing monitors were visible to the participants. This was in some respects a special case.

At Reading each of the neighborhood communication centers was originally provided with two view-finder cameras and a small switching center on the pattern of standard, simple, one-way TV production centers. At each of these centers three small monitors were provided for the production people, one for each camera and one to show the outgoing "program line." The expectation was that a technical aide, and later the users themselves, would operate the switching center, cutting between cameras in what have been described as "crossfire viewing angles." However, the senior citizen users showed no interest in performing technical duties such as camera handling and switching, and soon the second camera was modified for use as a vertical graphics camera, leaving only one conference camera operated by the technical aide. The local switching center with the three small monitors then lost some of its functions. The first of these small monitors became redundant since the camera had its own view finder. The monitor for the second camera continued to be needed because this camera had been mounted overhead, most likely with the view finder removed. The "program line" monitor had little use except at the moments when the graphics camera was switched on or off the line, if only to assure the aide that the switch had actually taken place.

In the particular situation at Reading it was probably best that there was no outgoing monitor visible to the participants. The cue that a person took to start talking was the appearance of his or her picture on the incoming monitor, which could take place only if that particular center was feeding the common downstream line. It would have been confusing to have another monitor showing the person's picture before it was actually "on the air."

It has been suggested that the constant presence of self-view on a local monitor might be expected to increase self-consciousness (Baker, 1978). This may be so, but the question is, does it then have a negative impact on the amount or quality of interaction that takes place? Maybe it's little different from looking in a mirror. We can, however, ask if the self-view monitor is actually needed at all. It is certainly not a part of the familiar face-to-face situation. It may be no more than a vestigial remnant of the one-way TV control room in which the most important monitor is the screen that shows the outgoing program. If so, it is like the socket for a buggy whip that was still part of the dashboard of early automobiles.

However, some of the self-operated systems provided by the common carriers have multiple picture sources, and a local monitor can serve the traditional "program monitor" function by showing everyone just which source is on the line. In the case of the Picturephone service there are seven cameras, four of which are switched automatically, which together then constitute one of the four video sources that are switched manually. The participant (usually the chairperson) who operates the switching control panel in the table needs to know what camera is on the line, and by some more obvious means than a lighted button on a control panel.

There is also the condition in which some graphic representation, slide, or videotape is the subject of discussion, when it would need to be seen equally well at both ends of the line. It may have been the anticipation of this situation that induced the designers of common carrier systems to put an outgoing monitor of equal size next to the incoming monitor.

None of this says anything about placement of this outgoing monitor. Certainly it need not be close to the camera; eye contact with oneself is not required. It can be below, above, or off to the side. It could be very small and built into the desk top. The outgoing picture could be displayed as a small inset at the bottom or corner of the incoming picture screen. Since twelve or fifteen screen widths would not be too great a viewing distance for this self-view image, if inserted into an image viewed from three screen widths it would occupy only one-ninth of the screen area or less.

Still another possibility is a kind of time-division multiplexing. The chairperson could have, among his or her controls, a button that displays the local image full frame on the central screen. This would be a good way to display a locally originated graphic so that all local participants could see it. If this button were spring loaded, the chairperson would be inclined to use it only briefly, except during the discussion of a graphic or the running of a videotape, when the button could be locked down.

Preview monitors, in television production, are for the purpose of checking a picture before it is put on the air. When there is only one camera at a location there is no need for a preview monitor, since the camera is never off the line. Previewing assures the producer that the camera is on the right subject, properly composed, focused, and the like. It provides engineers with the opportunity to make technical adjustments at the last minute that might otherwise have to be seen by the audience.

In interactive systems the equipment is reliable enough, and technical quality standards are flexible enough, that technical adjustments are rarely needed during a session. In the common carrier systems in which the cameras are not manned, control over the content of a shot is usually exerted over the subject in front of the camera rather than through camera itself. Participants are seated in specific places. If easel graphics are used, the easel is attached to a wall, within the view of a fixed camera.

A good example of this is the fixed overhead graphics camera. A camera operator in a TV studio moves the camera to perform the left-right panning movement and the up-down tilting movement. In the case of the overhead camera, these movements are accomplished by the graphics operator, who moves the artwork left-right or toward-away under the fixed camera. (The only camera control is zooming, with which the operator may increase or decrease the field of view the camera covers.)

In the AT&T meeting rooms, where the graphic stand and camera are off to the side of the room, two small monitors are built into the stand (see Figure 9, in Chapter 1). One of these shows the output of the graphics camera at all times and may be used for previewing that camera before it goes on the line, or operation of the graphics while it is actually in use. The second small monitor is a duplication of the main incoming picture monitor and allows the graphics operator to see the distant participants and interact with them if he or she wishes. A person handling graphics would be about four screen widths from either of these small monitors. It is necessary that a person who is presenting a graphic in this manner — pointing to specific items as they are discussed, moving the graphic materials up or down, left or right, zooming the lens in or out when necessary, and sometimes several of these at once — do so by watching the monitor, not by looking at the graphic itself. Only by watching the monitor can he or she know what is in the picture, and whether it is in sharp focus. In this regard it is essential that the monitor and the graphic are similarly oriented. That is, they should be situated so that the operator need only raise (or lower) his or her eyes to see the monitor, without having to look off to the side. A monitor placed in any but the same direction as the graphic materials to be manipulated will require a constant translation of direction on the part of the operator — a desire to move north on the monitor will require an eastern movement of the graphic, and so on — a trick as difficult to perform as patting your head and rubbing your stomach at the same time. These principles were discovered in relation to the first overhead TV graphic camera arrangement developed by one of the authors for classroom audiovisual use at UCLA in 1959.

The General, Overall Shot

Another issue that must be resolved in designing a video system is the question of the overall shot. The theory behind the use of this shot is that (1) people like to know the spatial relationships between the individuals that are seen on the screen, and (2) a good time to orient them to this is when there is no particular speaker to be featured, or no particular listener. The Bell System Picturephone Meeting Service systems, both in the United States and Canada, use voice-actuated switching between conference cameras, and switch automatically to a camera showing an overall shot if no one speaks for a set number of seconds (settings of four to twenty seconds have been used).

The need for the overall shot is often questioned. One reason for this is that in such shots individual faces are quite small. When the shot includes 6 people, as in the Picturephone system, or 9, as in the experimental Bell Labs studio, the size of people in the image puts them at a tremendous virtual distance. In a typical common carrier studio with a viewing distance of 5.5 W (such as in AT&T's Picturephone Meeting Service), a shot containing six people puts them at a virtual distance of 56 feet (almost 17 meters), well into far public distance. It is little wonder that the overall shot in the Los Angeles Meeting Service studio is not considered to be very valuable. In the Chicago Picturephone facility, where a composite shot of the images produced by the three conference cameras is used (see Chapter 1), the heads are all about 15 percent of screen height. Virtual distance is then reduced to 36 feet (about 11 meters), still in far public distance but a marked improvement over the standard way of showing all 6 people at once on the screen.[4] A screen at 3 W would put the people in this composite shot at a distance of 20 feet (just under 6 meters), about the middle of near public distance. Not close enough to one's listeners for one to feel comfortable in casual conversation, but close enough to read gross facial expressions.

It must be remembered that when a 2-shot is used for the speaker in the downstream direction and a shot of 6 people is coming back, it is not a symmetrical relationship. A Los Angeles speaker, picked up on a conference camera in 2-shot and displayed in New York will be seen at a virtual distance of 22 feet (6.8 meters), while the 6-person shot returned to Los Angeles shows the speaker his or her listeners at a distance of 56 feet (17 meters), apparently 2.5 times as far away. Both 22 and 56 feet are far public distance, and it may not make too much difference in this case; speakers at both ends may tend to project their voices and the listeners may expect them to. If everyone were within 3 W of the screen the virtual distances would be reduced to about half of what they are: The speaker would be talking to people who seem to be 28 feet (8.5 meters) away (between near and far public distance), while the viewers would be looking at a speaker less than 11 feet (3.5 meters) away (far social distance). This might unbalance things more.

If the general shot is used only when the people in it are listening, with the purpose of providing visual feedback to the speaker who is talking to their image, he must be able to see their expressions, especially the subtle movements of the muscles around the eyes. If the general shot puts people at such great virtual distances that subtle facial expressions are not really visible, then why is it used at all?

This shot may also be a holdover from one-way television. The standard system used by most TV directors when a television program is not scripted is to use one (or more) cameras on close-ups of people with another camera always ready with a wider-angle lens on what is called a "cover shot." Then, whatever happens, the director can always "go to cover," and will have a shot that contains the action. If a person speaks at the other end of the table, or someone suddenly rises to go to the chalkboard, the cover shot will keep the subject in the frame. Then the close-up camera, off the line, can swing to the new subject and the director can go back close again. This same concept also goes back to the traditions of film. In film a sequence of shots, to be totally understandable to the viewer, must traditionally contain what is called in film the "establishing" or "orientation" shot. This is a long shot or cover shot showing all relevant locations in the scene and their spatial relationships. There was a time when all films were shot in overall view only; the close-up had to be invented.

Is there a need for orientation — for establishing the spatial relationships of people at a distant location? The only relationships that exist, usually, concern who is sitting to the left or right of whom, but it is comfortable if this can be kept straight. If in a two-shot a person is seen to turn and speak to another participant at the table who is out of frame, it would be nice to at least have a general idea to whom he or she is talking.

Cutting Problems

Television talk shows generally manage to keep their audiences well oriented through their use of cover shots and camera manipulation. In a teleconference system where switching is done automatically between three two-shots, for instance, the following confusing sequence *can* occur.

Say you are watching a two-shot and the person on the left of the shot turns and speaks to someone outside camera range. There is a sudden automatic cut and you now see another two people, but here the person on the right of the screen is turning and talking to someone outside the shot on the *right.* Section b of Figure 33 illustrates the composition of these two shots. When you simulate the effect of the cut by looking quickly from one to the other, you will get the feeling that the two people are looking in opposite directions, not at each other but away from each other, and must therefore be talking to two other people. In fact, however, they are seated side by side and talking to each other. The illusory effect is caused by the fact that when a picture changes on the screen your

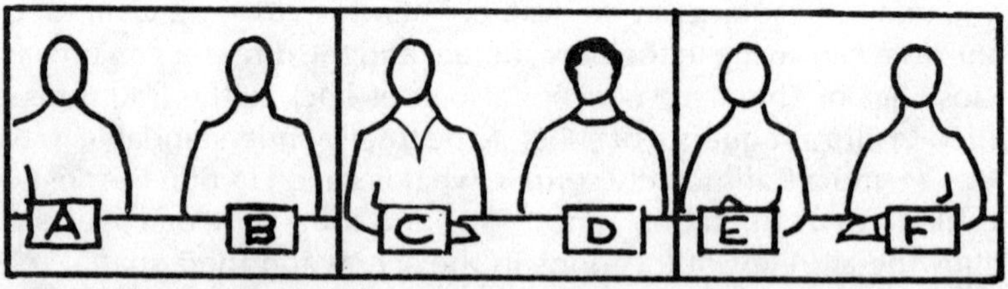

a. The three shots to which automatic switching systems are generally limited.

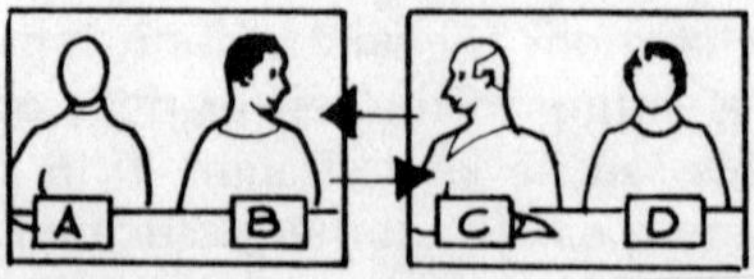

b. When B talks with C, the cut between the two shots can cause confusion. When shot A-B is on the screen, your eye is on B's face. When C replies and shot C-D appears, you are left focused on passive D and must come upon C from behind him. Hence the juxtaposition gives the impression that B and C are looking away from each other.

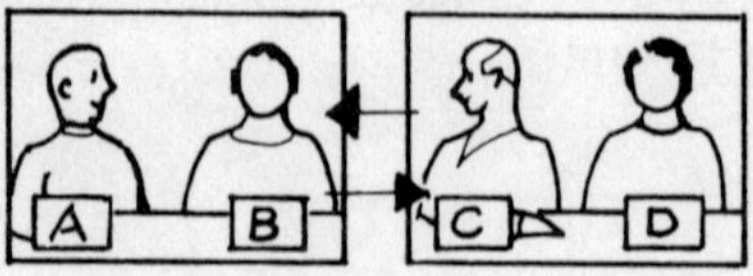

c. When A talks with C, they are in the same area of the screen and the cut does not cause such confusion.

d. When A talks with D, you must look back and forth in the natural direction after each cut and the spatial relation is perfectly clear.

e. Table of combinations of people speaking to each other and the shots that are automatically transmitted. Of the fifteen possibilities, only three are really bad juxtapositions; six are acceptable (as in the case of c, above, and six are good, as in the case of d).

Interaction	*Automatic Cutting*	*Evaluation*
A talks with B	AB is held	good
A talks with C	AB intercut with CD	acceptable (see c)
A talks with D		good (see d)
A talks with E	AB intercut with EF	acceptable
A talks with F		good
B talks with C	AB intercut with CD	bad (see b)
B talks with D	AB intercut with CD	acceptable
B talks with E	AB intercut with EF	bad
B talks with F	AB intercut with EF	acceptable
C talks with D	CD is held	good
C talks with E	CD intercut with EF	acceptable
C talks with F	CD intercut with EF	good
D talks with E	CD intercut with EF	bad
D talks with F	CD intercut with EF	acceptable
E talks with F	EF is held	good

Figure 33: Orientation.

eye is left on the particular area of the screen you were last watching, in this case the face of the first speaker. After the cut you find yourself looking at an impassive person and move your attention quickly to the active speaker, but you come on him or her from behind, so to speak. You have looked from the back of one person's head to the back of another's, hence you feel they must be looking away from each other.

If both people who speak happen to be on the same side of the frame (section c, Figure 33), the illusion that they are looking apart is not as strong. Only if one outer person of the four is talking to the other outer person, across the two persons between (section d) are these two shots sure to express the correct spatial relations.

A TV director, or a camera operator on his or her own, will avoid the problem shown in section b of the figure by simply panning the camera one person along to hold the two that are in conversation. If they are separated by others, as, for instance, if participant B is talking to E rather than C, the camera may widen out to show four people. In a system that uses fixed cameras without manual controls, this cannot easily be done.

According to the values of one-way TV production, anything that causes disorientation, even if only momentary, is considered poor technique. It is also disturbing in interactive communication, but it occurs only when the interaction is taking place between participants at a single location. At such times the distant participants are watching an essentially one-way presentation. Because it is rare, in the design of self-operated or automated systems it has been disregarded. Among six participants there are fifteen different combinations of two people who could engage in conversation (section e of Figure 33). Assuming a system limited to three two-shots containing A + B, C + D, and E + F, only three of the fifteen combinations could produce the confusion shown in section b of the figure. These combinations are: B conversing with C, B with E, and D with E.[5] Six combinations present two conversants on the same side of the screen looking in different directions (not spatially confusing) and six combinations keep everything perfectly clear. The situation of interaction within a group at a single location is itself not very common, so a system might be expected to operate for months without encountering a really disorienting juxtaposition of shots.

The participants in the systems just described have been assumed to be in a line or semicircle before the cameras. A quite different placement was experimented with by Bell Labs in which six participants were seated in a circle. Three cameras covered this group from three sides, each covering three participants. Orientation was almost impossible. In conversing via the TV system with persons in this particular meeting room, it was the experience of one of the authors that despite the occasional general view he really had no idea who was seated where. He finally ignored this uneasiness, convincing himself that such information was irrelevant to the content of the discussion, and contented himself with simply knowing who he was talking to at each moment. We imagine that a similar feeling of orientation blindness accompanied the use of the first-generation Picturephone service (see Chapter 1).

In summary, it can be said that spatial orientation is a kind of natural need. We feel uncomfortable if we do not know the environment we are in, and, to the extent that we share the distant space of another teleconference group, we are more comfortable knowing the spatial relationships there. This is usually limited to spatial relations between people, but could also include graphics, stands, easels, or other locations if people move to them during a conference. Because most systems place their viewing screens at five screen widths or more from the participants, faces are small and people are hard to recognize. The usual overall shot is not very successful in achieving orientation, and even less so in providing visual feedback to a speaker. This could be different if screens were placed closer to participants. Orientation is a problem only in systems such as those operated by the common carriers, in which operation is highly simplified through the use of fixed cameras, automated switching, and the like. Where cameras are manually operated and have zoom lenses, a continuous shot, changing subject through panning, zooming, and so on, can both establish and retain orientation.

notes

1. The same may be done with screen heights, giving a distance figure that is always 4/3 that of distance measured in screen width. Ratio of TV image height to width is 3:4. One screen width thus equals one and one-third screen heights, so the number of screen widths multiplied by 4/3 yields number of screen heights.

2. Stapley's experiment was performed as follows: The two objects (a man and a picture of the man) were not placed side by side, but 90 degrees apart. When the subject of the experiment looked ahead, he saw a TV screen carrying the image of a face. The image cast on the subject's retina was the size that the real person's face would create from a distance of 100 inches (virtual distance = 100 inches), although the TV set was much closer (38 inches). When the subject looked to his right, he saw the real person some distance away, who then moved slowly toward him. When the live face appeared to be the same size as the TV face, the subject would say "Now." The same thing was done with the person moving away from the subject. The live person carried a model of the TV monitor frame surrounding his face, which if he had stopped at 100 inches would have duplicated in retinal image size the TV monitor and the face on its screen. However, the average subject of the 21 that took part felt that the live person was matched in size when he was 220 inches distant — over twice as far away.

One subject extended his thumb at arm's length, like a landscape artist, to gauge the relative sizes. With this aid he was evidently able to concentrate only on the factor of retinal image size; he placed the experimenter at 103 inches. Because of this he was excluded from the analysis. Further experimentation is clearly needed here to determine just what factors were causing the unexpected results.

3. When looking at a two-shot with one person on either side of the screen, a viewer's eyes would seldom be on the very center of the screen, where the camera lens is located. At the very worst, a person might be watching something at the side of the screen (a speaker who is leaning out, perhaps). Maximum possible displacement might be as much as half the screen width. At 3 W this would cause a maximum parallax of 9.5 degrees, hardly enough to be noticeable.

4. Calculation of these figures is as follows: Starting with the relationship (head height on screen)/(viewing distance) = (real head height)/(virtual distance), we have virtual distance = (real head height × TV viewing distance)/(head height on TV screen). In the Los Angeles Picturephone meeting room, viewing distance is 5.5 W, screen height is 16 inches (40 cm), and heads are 10 percent of screen height in the overall shot of 6 people and 24 percent in the 2-shot. Average real head height is 9 inches (23 cm).

5. People sitting side by side at conferences frequently engage in *private* conversations, but since these are not intended to be heard by all, they are conducted sotto voce — they do not concern viewers at the other end of the line, and do not actuate voice-operated switching, so they are not relevant to this discussion.

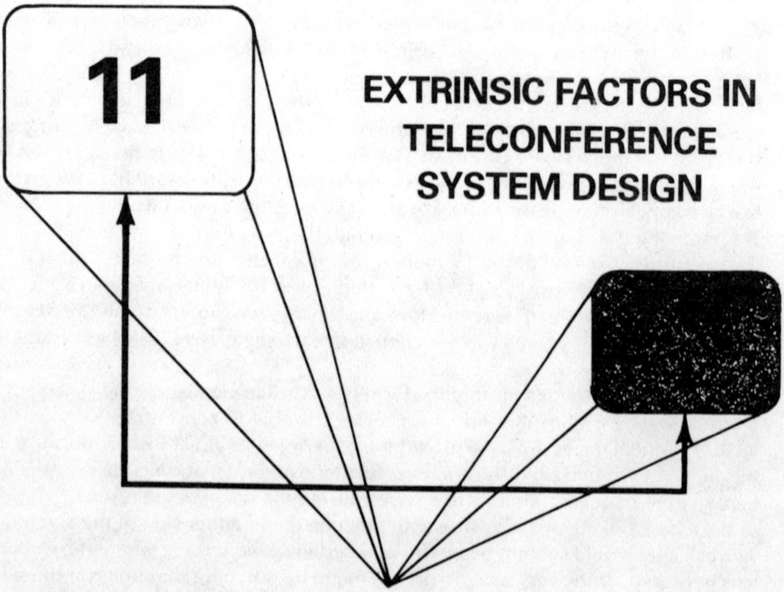

11 EXTRINSIC FACTORS IN TELECONFERENCE SYSTEM DESIGN

Success or failure of a teleconference system more often than not, it seems, will rest on some factor that is no part at all of the system itself, but rather of the situation or set of circumstances surrounding the project. In some cases the failure of the system to fit its situation can be laid at the door of its designer, or at least the administrators of the project, for not understanding the situation better. Sometimes it seems there was nothing anyone could do. Many of these problems are common to any kind of innovation. Since this is a book about media, not the acceptance of innovation, it is not the place for a thorough analysis of these extrinsic factors. All we will try to do here is to recognize and identify some major problems and warn that they too must be anticipated and dealt with if an interactive media system is to be successful.

User Acceptance

The solution of the problems in this area must be begun long before the system is installed, preferably before any major decisions are made concerning its characteristics, what media will be used, where the terminals will be located, and the like.

It appears to be absolutely essential that management of the organization that will use the system perceive a need for the system from the very beginning. This perception of need, when strongly felt by management, will be transmitted downward to subordinates and eventual direct users far better than a need perceived by underlings can be transmitted upward. Of course, management and "labor" often have different objectives. Management may be concerned with enhancing prestige (of one department versus another or one firm versus its competitors); it may be concerned with saving money or increasing efficiency. The direct users may be more concerned with convenience, improvement of individual effectiveness, and the like. If the staff people feel the new system is a threat to the travel that they enjoy — to conventions and other meetings — they may misuse or underuse the system, even without fully understanding why.

Some media systems are expensive, especially television systems, and often must be jointly used by several organizations in order to be financially feasible. It appears that the problems caused by such joint use become greater the more closely related the user organizations are. Two departments within an organization, for example, may not be able to cooperate on such a thing at all. If there can be only one terminal facility, for example, it must go somewhere, its staff must be responsible to some manager, its expenses must come from some budget. It is difficult for any of these functions to be entirely neutral. If the teleconference studio is located in department A's area, it is on their "turf," so to speak. It will tend to be operated by A personnel, oriented to A's needs and priorities. At the least, department B will overlook many of the opportunities the system offers; at the worst B might reject it entirely.

Convenience seems to be an important factor in user acceptance. A terminal that is located on some other organization's turf is psychologically inconvenient, no matter how close it may actually be. Lack of physical proximity is a major inconvenience and may be responsible more than any other factor for the slow growth in the use of public (common carrier) systems where there are only one or two terminals in a city. The time involved in leaving a building, traveling across town, entering another office, and so forth may be relatively short when compared to traveling to a distant city, but some users have actually said that the two are equally inconvenient. Not having a telephone on one's desk, for instance, and having to cross the room to use one is considered such an inconvenience as to be practically unthinkable in the modern world, even in many developing countries.

In addition to psychological inconvenience and physical inconvenience, there is temporal inconvenience. Is the terminal available when you want to use it? How far in advance must you schedule its use? Who can preempt it from you? Even when these problems are no greater than they are in planning ordinary meetings, they look larger when they occur in connection with a teleconferencing system the perceived purpose of which is to eliminate such constraints.

Operational Support

Operational support is another extrinsic factor that can make a big difference in the success of a telecommunication system. Adequate funds for maintenance, including personnel, are a large part of this. The important thing is that the system be reliable and consistent. Occasional breakdown of parts of a system (such as the loss of one camera out of three, for instance), while an inconvenience, can be overlooked as long as such events are infrequent. When catastrophic breakdowns that render the system unusable (complete loss of audio transmission, for instance) are frequent, it can become impossible for users to incorporate the system into their scheduled activities.

Most organizations that have built private telecommunication systems employ a chief engineer and a number of technicians depending on the size and geographical spread of the system. If the technical staff is adequate, it will be possible for them to do "preventive maintenance" — testing, checking, overhauling equipment on a fixed schedule — so that breakdowns tend not to occur. If this is not possible, maintenance may have to be handled on a crisis basis, waiting until something actually breaks down before it is given attention. In commercial broadcasting and in common carrier operation, where loss of service means immediate loss of income, the cost of duplicate equipment for backup purposes can be justified. In private systems this is rarely possible. The AT&T Picturephone Meeting Service is *already* using backup equipment for its long-distance transmission. AT&T must maintain backup video lines and routes between most U.S. cities for immediate use in case of equipment or power failure in their primary commercial service. Since these backup lines are normally merely kept in reserve, they are available on a secondary basis for the meeting service, but can be preempted on a moment's notice if needed.

Equipment of any kind has a fairly fixed lifetime, which can be shortened by abuse or poor maintenance and lengthened by proper care and replacement of worn-out parts. Failure rate, which can be greatly lessened by preventive maintenance, varies in a peculiar fashion over the lifetime of the equipment. It starts high, lowers to a minimum fairly quickly and finally toward the end begins again to rise (see Figure 34).

At the beginning, and for the first 200 hours or so of use, the failure rate will be very high, many times what the ongoing rate will be once everything is well broken in and stabilized. For this reason new systems should be operated for testing, training, experimental purposes, or for no purpose more than to run off 200 hours. The military has long insisted that all electronic equipment purchased have its first 200 hours already "burned in." After the initial "bugs" have been discovered and eliminated from the system and it can be put into operation, the failure rate will continue to go down until about 2000 hours of use. During this period maintenance needs will be high. More technician time will be required (perhaps two or three times as much) than will be needed later on. It is a mistake to try to get past this period with the same limited personnel that is planned for the subsequent stable period of service, and it is important that initial impressions be good ones. A system establishes its reputation when it is new. Management should not pressure system personnel into starting operation too soon and, to prevent this, a contingency period should always be allowed following installation, for unexpected delays.

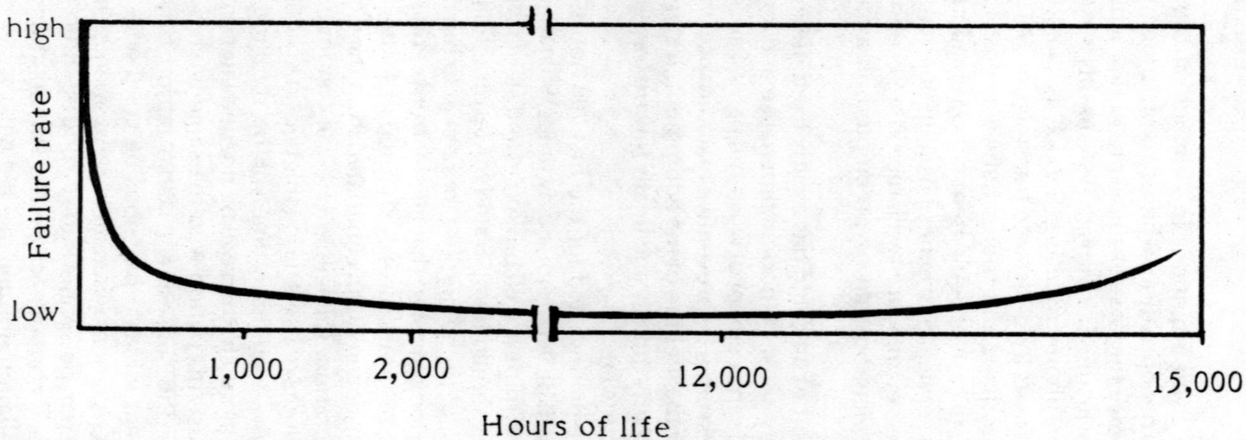

Figure 34: The "bathtub" curve.

After 2000 hours or so, the failure rate stabilizes at its lowest point, which it then maintains until the equipment reaches the age at which components begin to wear out. This can begin at 15,000 to 20,000 hours or not until much later, depending on the kind and quality of equipment. At 8 hours a day, 240 days a year, the 15,000 hour level is reached after 7 or 8 years. Then the failure rate begins to go up again, doubling and tripling quickly. Soon the time and parts required to keep the equipment functioning becomes more costly than the replacement of equipment. Some or all of the equipment is replaced and the cycle begins again.

This whole phenomenon is shown in the graph in Figure 34. Because of its shape, this is known as the "bathtub curve." Over the years the bathtub has grown somewhat shallower and longer as better equipment — transistors replacing tubes, for instance — has reduced the shakedown period at the beginning and extended equipment lifetime at the end. Some organizations have contracted with outside firms for maintenance services at the beginning to cope with the higher failure rate. This generally makes more sense than overstaffing at the beginning. It is important, however, that such outside maintenance contracts include response-time clauses. The purpose of such service should not be forgotten — to make the system reliable — and a maintenance service that takes several days to respond to a trouble call is not very much help at all. Even if a 24-hour-or-less response-time clause with penalty costs more, it may well be worth it, especially at the beginning when user acceptance is at stake.

Substitutability of Teleconferencing for Travel

There are many cases in which an innovation is accepted and undertaken by an organization under the expectation that it will help to solve some particular organizational problem, such as operational costs, and, despite obvious success in other directions, fails to solve the problem for which it was created, and is then considered unsuccessful.

One example of this is the saving of travel costs by substituting teleconferencing. Organizations often see the possibility of substantial savings over sending people on long intercity trips in which high travel and per diem expenses are incurred plus the "opportunity cost" of taking people out of their normal productive activities for the time of the trip. Many people in this field anticipate that the future use of telecommunication (beyond telephone and letter correspondence) will save a large share of daily commuting within an urban or metropolitan area. Some part of many office workers' duties could be, according to this thinking, performed at home or in local community work centers close to workers' homes, so that daily trips to the office would not be necessary. The more telecommunication that can be provided between these subsidiary workplaces and the central office, the less lost time and effort in fighting traffic, the fewer cars on the road, the less pollution, and the less overall expenditure of energy. In recent years the conservation of energy sources has become a prime consideration.

Sponsors of research, such as the National Science Foundation, the Federal Energy Administration, and the U.S. Department of Transportation, have funded many studies aimed at surveying possibilities and assessing feasibility of the substitution idea both in urban and interurban situations. In one of these studies, made by the National Academy of Engineering (Goldsmith, 1969), travel was divided into four kinds and estimates were made of how much of each kind could be replaced by telecommunications:

purpose of travel	percentage of such travel that can be replaced by telecommunication
earning a living	20
family business	20
educational and civil affairs	25
social and recreational activities	5

Goldsmith (1974) estimated that about 18 percent of "urban travel" could be replaced with telecommunication. This same figure was used five years earlier in the National Academy of Engineering study cited above, relative to "urban area vehicles miles," which in this day of limited public transport in America probably amounts to about the same thing as "urban travel."

In respect to the conservation of energy, Goldsmith (1974) estimated that transportation consumed 25 percent of all energy in the American economy, and that autos, which were used for 90 percent of all intercity travel, used 14 percent of all energy consumed (30 percent of all petroleum). Harkness (1973) estimated that 11 percent of the nation's use of petroleum went for commuting to work, and about one-fifth of that (2 percent) could be replaced by telecommunication. Another study, by the Department of Commerce (Anker-Johnson, 1974), estimated that as much as 5 percent of U.S. petroleum consumption could be saved if telecommunication could be substituted for travel to the greatest possible extent.

Dickson and Bowers (1973: 299) calculated that the cost of one person making an overnight trip from New York to Chicago for 6.5 hours of face-to-face meeting would pay for 75 hours of Picturephone interaction. Two persons making the same trip, it could be inferred, could have had about twice as much Picturephone time as this for the cost of travel. At the time of this study such intercity Picturephone service did not exist; the authors assumed the rates to be ten times those of the then current audio telephone rates. In fact, market trials during the early 1970s were so discouraging that Bell shelved the whole idea of a video telephone service, at least for the time being.[1]

While we have an extensive literature of a speculative nature on the feasibility of substituting telecommunication for travel, we have very little in the way of carefully documented experience or experimental testing of the replacement hypothesis. Occasionally an organization will report that 60 or 80 percent of its previous trips were replaced, but will fail to convert trips into dollars and make any kind of cost comparison. A NASA report was quoted as indicating that $684,000 was "saved" in one 9-month period after travel was replaced with an audiovisual system (Hough and Panko, 1977: 190). However, this did not take into account that at least some of the "saved" trips might never have been authorized or taken at all if the more expensive travel mode had been necessary.

In general it can be said, however, that while hard data on replacement savings may be difficult to come by, the sum total of all communication within or between organizations can be shown to rise when teleconferencing systems are available, especially if the various extrinsic factors of the situation are favorable. In many cases, when this happens, the requirements for face-to-face conferencing tend to rise as well, so travel budgets may actually increase while travel is being "replaced." Very complex cost/benefit studies analyzing the full organization would be necessary before it would be possible to put a monetary value on increased communication. The obvious question — how much an organization can save by replacing travel with this or that kind of teleconferencing system — is evidently too narrow and simplistic to elicit a useful answer.

It may be concluded, then, from the experience of organizations that have gone seriously into interactive communication, that although many meetings that would have been held face to face are transferred to a telecommunication medium, a much larger number appear to have been *generated* — meetings that would not have taken place at all were it not for the availability of the medium. There are many examples of this in fields such as education, health care, and industry. The Quebec and the Wisconsin uses of telephone and other media provide isolated persons with opportunities for education and training in many subject areas. Telemedicine allows specialists in large hospitals to "meet" patients in neighborhood clinics or isolated communities and diagnose and treat ailments, with the most efficient use of time and energy. Short et al. (1976) report:

> In the last ten years at least 24 telemedicine projects have been instituted or planned in the United States or Canada; all have the general goal of delivering health care to as many persons as effectively possible.

A classic example is used by the Bell Telephone System in their promotional literature. A salesman, the phone company states, can make thirty "visits" per day if he uses the phone, as opposed to the three to which he is limited if he makes appointments and sees people in person. Assuming he is only one-quarter as successful by phone, he can still do twice as much business.

The generation of more communication overall as a result of using media may also be due to several kinds of learning from experience. Short et al. (1976: 167) identify five possible kinds of such effects:

(1) Individuals learn to communicate more smoothly over new media.

(2) Individuals learn to determine more accurately which communications media are effective for which meetings.

(3) Small groups learn how to organize themselves, perhaps by special chairmanship arrangements, to use new media more effectively.

(4) Organizations learn more about the capabilities of new telecommunications devices, perhaps by innovations diffusing from organization to organization.

(5) Organizations learn how to reorganize their internal structure and work flows so as to use the new telecommunications possibilities more effectively.

Short et al. note the need for further research in this area, however; what data there are seem to indicate that initial attitudes and behavior habits are highly resistant to change from this new learning. This may be merely because most research to date has been on new systems with new users; there have been few systems in which very many people have gained the experience of extended use.

The bulk of the research on interactive communication has been done by the Communications Study Group, supported mainly by the British Post Office. The main thrust of this work has been to attack the problem of estimating how much of which kinds of face-to-face communication might be transferred to which kinds of media systems. Chapter 12 summarizes the study group's findings.

note

1. Between 1935 and 1938, according to AT&T, the first videophone service was operated by the German Post Office between videophone centers in four cities: Berlin, Leipzig, Nuremburg, and Hamburg. Other reports describe a Berlin-Munich system.

12 WHAT MEDIUM FOR WHAT PURPOSE?

An important issue in interactive communications, one that has generated a great deal of research, involves such questions as: Under what circumstances should a given medium be used? What can be done best with an audio medium? When should video be used? What kinds of things can be done only in the face-to-face situation?

In the light of the large variety of features that both audio and video media can possess, it is obvious that any question involving audio or video as a class is necessarily vague and can only generate the most gross information. If there are, say, a dozen different kinds of audio systems, differing in the kinds of visual augmentation provided, degree of automation, and the like, there is no certainty that determinations of effectiveness for one kind of audio system can be applied to another. The best we have been able to do is to study the few systems that we have, as a whole, without attempting to relate specific features to specific results.

Researchers realized that it was of immediate importance to evaluate existing systems in the field. Such studies were generally investigations of user attitudes as expressed in interviews or selected-response questionnaires (for example, Confravision was: pleasant/unpleasant, friendly/unfriendly, and so on, or statements for agree/disagree response on a simple seven-step scale). A very good review of these studies, especially those by the Communications Study Group (CSG), University College, London, is provided in Hough and Panko (1977). What people say they feel about a system is not always in accord with what they actually do about it, a phenomenon that Hough and Panko recognize. This may be because they are thinking only of the intrinsic factors of a system when responding to the questionnaire; when it comes to action the extrinsic factors begin to enter in (convenience, ownership, rewards, and so on). Hence the value of such attitude studies alone as a predictor of the popularity of future systems is limited.

Hough and Panko provide a comparison of four such attitude studies, concerning the following systems: the University of Quebec audio teleconference system (briefly described in Chapter 2), the Bell Canada video system (similar to the AT&T second-generation Picturephone meeting service), the Remote Meeting Table audio system (described in Chapter 3), and the British Confravision video system (described in Chapter 1). Opinions on the effectiveness of these systems for various task purposes were obtained from a total of 700 users, from 140 to 193 per system. In each case evaluations were made on a scale of 1 (completely satisfactory) to 7 (completely unsatisfactory). For the summary given below we have converted these scores to percentages of complete satisfaction.

Twelve activities that take place in teleconferencing are listed in Table 5 (we have called these "tasks" or "purposes" above). About half of these can be described as personal activities, in which individual personalities and character traits play an important role, and the other half can be called impersonal, in which such personal factors are unimportant. They are listed in the order of their average satisfaction scores, that is, those activities that can be most successfully pursued via teleconferencing are listed first. It is interesting to note that this order is almost the same in each of the studies.

The range among the studies is small where impersonal activities are concerned, but two or three times as great when it comes to the personal activities. The two video systems, taken together (see the bottom line of Table 5) averaged 8 percentage points above the audio systems on personal activities (40 to 48, or a

TABLE 5 Average Satisfaction Scores (in percentages)

	Audio Systems		*Video Systems*		
	University of Quebec	*RMT*	*Bell Canada*	*Confra-vision*	*Total*
Number of respondents	177	140	190	193	700
Impersonal activities					averages
Asking questions	71	*	71	73	72
Exchanging information	77	63	69	71	70
Exchanging opinions	70	*	67	73	70
Giving or receiving orders	57	64	61	66	62
Problem solving	56		60	61	59
Decision making	61	56	59	59	59
Personal activities					
Maintaining friendly relations	40	*	56	64	53
Resolving disagreements	40	51	47	54	48
Getting others on one's side in an argument	41	*	43	50	45
Persuading	43	44	43	49	45
Bargaining	33	44	41	49	42
Getting to know someone	26	27	36	43	33
General averages					
Overall	50	50	54	59	53
Impersonal	65	61	64	67	64
Personal	37	43	45	51	44

	Audio Systems	Video Systems	Difference
Overall	50	56	12
Impersonal	63	65	3
Personal	40	48	20

*Several of the categories included in the other surveys were lacking in the RMT study.

difference of 20 percent). In regard to the impersonal tasks, the difference was only 2 percentage points, or about a 3 percent difference.

This evidence of little difference between audio and video systems on the impersonal tasks (which generally account for at least 75 percent of business meetings) has been taken as proof that audio is the equal of video for these purposes. Another way of saying this is that impersonal tasks are not sensitive to the effects of media. It is not known, however, if there were any single factors within these particular media systems to which impersonal activities *were* sensitive, or to which factors personal activities were insensitive.

One might infer from this table that 100 percent (fully satisfactory) in the case of each activity would characterize face-to-face communication, but such is definitely not the case. These studies were not comparisons with face-to-face interaction, and there is no certainty that face-to-face methods, if evaluated, would have fared better, or even as well. Few people would rate the average face-to-face meeting, as generally managed, fully satisfactory for any task, although individual meetings or those managed by particular techniques might receive high ratings. In another similar study, Confravision and face-to-face meetings were compared against much the same list of activities; in all personal activities face-to-face meetings received a higher rating (as much as 95 to 5 on "getting to know someone"), but in the impersonal activities Confravision was considered *more* satisfactory (as much as 81 to 20 on information exchange; see Hough and Panko, 1977: 142).

Having thus obtained a general idea of the effectiveness of teleconferencing in various activities, CSG researchers wanted to find out how much of the communication normally taking place in meetings and conferences would be amenable to teleconferencing. To do this they had to first study the content of face-to-face meetings. Then, by relating media type to meeting type, they might arrive at some kind of estimate as to how much conferencing would turn out satisfactory if done via audio means, how much would require video, and how much could not be done by teleconferencing at all.

The CSG conducted several extensive studies along this line, involving, all told, some 8000 meetings (Hough and Panko, 1977:

99). The results of one of these studies, in which almost 1800 meetings were reported, are shown in Figure 35, which lists twelve activities along with the percentage of the 1800 meetings in which each activity was reported to have taken place. The twelve activities in this list are slightly different from those in the Table 5, but this makes very little difference to the conclusions reached. We have divided the list, like the first one, into activities of an impersonal nature, relatively insensitive to effects of media, and those of a more personal, more sensitive nature. The classification is necessarily rough and somewhat arbitrary, since there can be affective elements in the most cognitive content. A person will often convey feelings, for example, about the information he or she is conveying. Policy decision making could be listed in either category, depending on such factors as the personalities of the people involved.

Figure 35 indicates only the percentage of meetings in which each activity was reported. There are no data on how much of a meeting was devoted to each activity reported for it—only that the task was involved. Naturally, many meetings contain more than one type of activity, so the percentages add up to far more than 100. It can be seen from this chart that the largest proportion of activities that took place at these meetings was impersonal in nature, and hence (according to the CSG research) not particularly affected by the medium (or lack of medium) used.

Further refinement of this analysis led to the identification of tasks that were commonly grouped together, leading to a classification of ten kinds of meetings, which the CSG called "clusters." Table 6 displays these clusters as vertical columns, and the meeting tasks or purposes are listed horizontally. An X indicates the inclusion of a given task in a particular cluster. The clusters were identified by letters A through J. They have been rearranged from their original alphabetical order here to display the manner in which they were then allocated to the three general "media": audio, video, and face-to-face. This was done largely on the basis of best judgment and consensus among the researchers.

The CSG, in our opinion, necessarily oversimplified the array of media possibilities in many of their studies, providing only two groups: "narrow-band" media (audio, telephone, telex, and so on) and "broad-band" media (television). Since there were very few

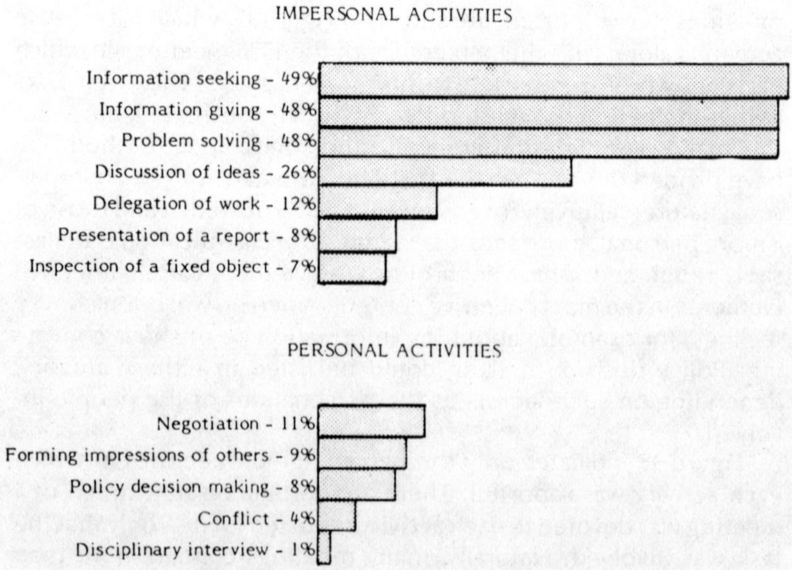

Figure 35: Activities reported in CSG 1973 Office Communication Survey of 1800 face-to-face meetings, by percentage of meetings reported to contain each activity (see Christie and Elton, 1975).

systems to study, any further breakdown would have been impractical. Future studies should be able to isolate individual factors and test, for example, the application of proxemics theory, which expects closer apparent distance to improve or increase personal communication.

In spite of their limitations, the CSG studies provide much knowledge about the effects that audio (in general) and video (in general) have on the communications they transmit. Primarily the studies indicate that in filtering out, so to speak, all visual means of expression, audio media are rendered less useful for communications that depend on what they call "social presence." On the other hand, and by the same token, they found audio media more effective than either audiovideo or face-to-face media for purposes such as policy decision making, implying that the interpersonal relations

TABLE 6 Meeting Tasks Clustered and Allocated to Media

allocated to:	*Audio*				*Video*			*Face-to-Face*		
clusters:	*A*	*E*	*H*	*J*	*C*	*D*	*I*	*B*	*F*	*G*
Meeting tasks (purposes)										
Policy decision making	X									
Information giving (to keep people in the picture)	X	X			X	X	X			
Information seeking	X	X	X	X				X	X	X
Delegation of work	X						X			
Discussion of ideas			X			X				
Problem solving		X	X	X	X				X	X
Inspection of a fixed object								X		
Presentation of a report									X	
Negotiation										X
Percentage of meetings classifiable in each cluster	7	11	9	15	9	9	8	10	5	6
Percentage allocated to each "medium"		42				26			21	
Percentage unallocated (because unclustered)		11								

NOTE: Other meeting tasks sometimes used in such studies, but not included above are: conflict resolution, disciplinary interview, forming impressions of others, getting to know others, making a good showing before the boss, and getting the issue settled quickly.

that are formed or conveyed by nonverbal cues may distract from or interfere with the success of these tasks.

Audio-only media eliminate many nonverbal visual cues, such as eye contact, facial expression, gesture, posture, proximity, and physical appearance. The CSG suggestion was that "non-verbal communication only has effects on the outcomes of mediated interactions to the extent that it determines feelings of social presence" (Short et al., 1976: 157).

Estimates were made concerning the suitability for audio and for audio-video in three different studies of face-to-face meetings, with results that were very much the same. Since this is only a quick overview, we have prepared an average of the three, as seen in Figure 36. The implication was that about two-thirds of all meetings could be held via teleconferencing, and two-thirds of these could be done by relatively simple, inexpensive, narrow-band media.

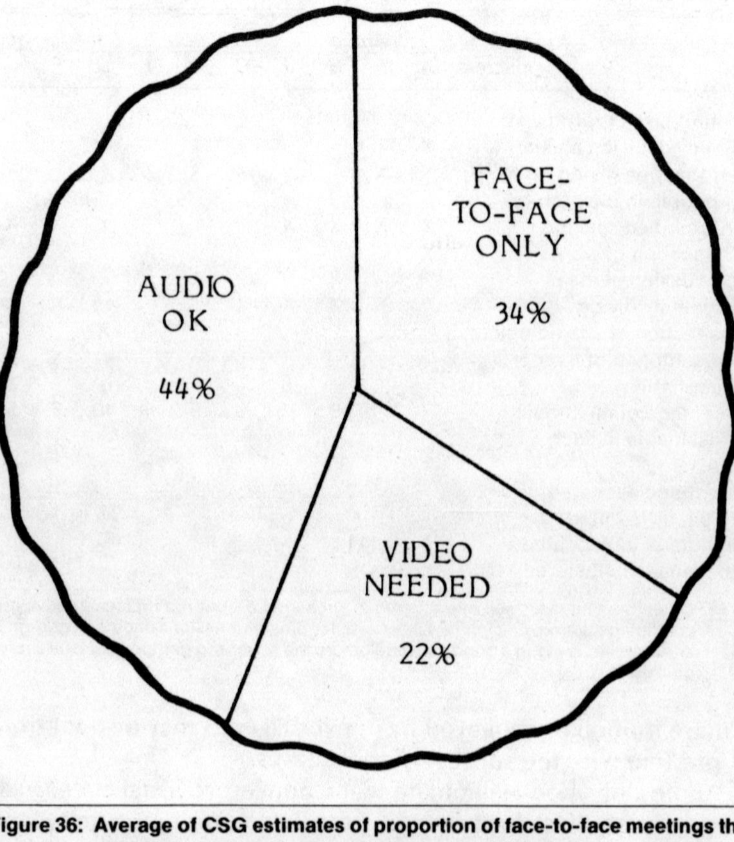

Figure 36: Average of CSG estimates of proportion of face-to-face meetings that could be conducted via audio or video conferencing.

It must be recognized that these CSG studies were carried on with the systems that were developed by the early 1970s. Confravision, for example, and the early Bell experimental system were perforce used to represent video systems in general. As the various characteristics of video systems are becoming better understood, more effective systems are being developed. Similar studies in the future may show a smaller proportion of meetings classed as possible only in the face-to-face mode.

The finding that for some purposes audio alone is superior to audio with still visuals, or to video, applies to cases in which the content of the meeting is purely cognitive and so abstract in nature that it can only be conveyed verbally. This suggests something that has also been characteristic of audiovisual one-way presentation: If visual elements are not relevant to the content, they can be distracting and can reduce the effectiveness of the communication. There are many cases in the CSG studies in which tasks were performed more quickly and effectively with audio-only systems. Further studies along this line may put even more meeting content into the audio-satisfactory category than the 44 percent shown in Figure 36.

13 DEVELOPMENTS TO COME

The pace of technological development seems to be constantly increasing; we are frequently surprised to find that developments that "those who know" told us were far into the distant future suddenly appear in workable form. Videotape is an example. In the early 1950s many firms were working on videotape, but it was said to be many years away. Then it suddenly appeared out of someone's garage. It had to be two inches wide and running at fifteen inches a second, but in twenty years it was down to half that width, and less than half the speed at better levels of quality, with one-quarter-inch equipment on the market for nonbroadcast uses, and a Japanese company demonstrating video recorded on a piece of cardboard.

Probably the greatest deterrent to the spread of full-motion video conferencing is the astronomical cost of video transmission. The PT&Ts of Britain and Australia, with their Confravision systems, and the Bell System, with its Picturephone Meeting Service, were able to provide services within a reasonable cost range, by using existing backup channels that were going unused, simply waiting in reserve for emergency need. Such backup channels are strictly limited, of course, and belong only to the large common carriers that handle network television traffic. Privately operated conferencing systems cannot afford the fees normally paid by broadcasters, and must find other means of distribution.

While satellite distribution would be practical in many cases, and highly applicable to organizations with many local offices throughout the country, another deterrent is evident — the current shortage of satellite channels. Even if satellites are spaced 2 degrees apart instead of the present 4 to 5 degrees, and they can all relay 24 channels, the number of channels is finite and competition for scarce channels is already strong. A natural solution would be to find some way to multiplex more than one TV transmission onto the usual TV broad-band channel.

The first step in this direction was taken when means for transmitting two TV channels on a single TV wave band was devised. This was done by transmitting only half as much information from each of the two multiplexed channels. Normal television broadcast standards in the United States call for a full picture (frame) every one-thirtieth of a second, each of which is divided into two fields of one-sixtieth of a second each, to reduce the effect of flicker. (One field contains all the even-numbered lines, the other all the odd-numbered lines, for a total of 525.) One method of multiplexing put one TV program on the even-numbered fields and the other on the uneven, dropping the intervening fields from each channel. The reception equipment filled in the missing field by duplicating the lines that were just received from the field before. Because only half the lines of each channel were transmitted, the picture quality tended to suffer. However, the definition is not necessary (or even desirable) in close shots of people, which is what most video conferencing consists of.

Another method of multiplexing is to drop out every other full frame from the transmitted pictures, and duplicate the intervening frames. With this method the picture quality does not suffer because all the scanning lines are transmitted; what does suffer is the effect of motion. But this is minimal, and in teleconferencing, where there is very little fast motion of any kind, the difference would probably never be detected. Except for sudden gestures and people quickly standing or sitting, eight frames a second would probably provide the illusion of full and smooth motion, and appear no different from full-motion TV at thirty frames a second.

With the introduction of digital encoding to television, a number of other ways of compressing the signal so it can be transmitted on a narrow-bandwidth channel are being developed. Some new terms are making an appearance in the vernacular: pulse code modulation (PCM), differential pulse code modulation (DPCM), motion compensating coding, and the device called the "Codec," that does the coding of outgoing signals and the *dec*oding of incoming signals, simultaneously if necessary.

It is the design of the codec that makes the difference between one compression method and another. Some of these methods break down the picture into areas that change and areas that do not change. In these systems only the "delta component" information, that is, the *difference* between one frame and the next, is transmitted; anything that is redundant from one frame to the next is supplied by the generative equipment at the receiving end on the basis of information already transmitted. If three-fourths of the frame is background and foreground areas of wall or table that do not change, only one-fourth of the frame needs to be transmitted. When the subject moves no more than to bat an eye or to move his or her lips, very little new transmission is necessary at all. In motion compensation coding, now in development, an even more sophisticated system is used. The receiver looks at the changes in a series of frames and predicts the motion of the subject in the next frames to come, which then become superfluous.

This is somewhat analogous to the "in-betweener" in the production of animated films. In the classical method the animator drew only the key positions of an action, say, every tenth or twentieth frame. Then another artist took those drawings, interpolated the actions between them and made the in-between drawings, usually one for every two frames.

One or two other examples from modern electronic technology may also be helpful here. The first is the character generator component of a modern CRT screen terminal. Instead of scanning a page of characters, as one might do when pointing a video camera at a typewritten page, and sending an image of each character, the characters can each be represented by a simple binary code (because characters belong to a finite set). Only the code need be transmitted; the character in all its detail is generated locally in the terminal.

In another case, a generative model is built in at the receiving end for a graphic representation, for instance, a bar chart. All that needs to be transmitted is a code to activate the model and the delta information to raise each bar from zero to the proper value, plus, of course, some labels for the x and y coordinates and identification for the various parameters.

The player piano is another analogue, this time from the preelectronic era. A recording device translates musical notes into punched holes in a wide paper roll. The positions of the holes determine the keys that are activated in the playback piano; the length of the holes determines how long the notes are sustained. Some of the later player piano models became quite sophisticated: Additional dimensions such as "attack" and "decay" were added, with the result that the playing of a master (Paderewsky and Rachmaninoff were among those so recorded) could be reproduced at another time and place, on another piano, with all of the technique, and imperfections, of the master himself. Because it was a piano that was reproducing the recording, the fidelity was perfect, limited only by the quality and condition of the playing piano. The *sounds* of the music were not recorded and transmitted to a distant place and time, but merely the *instructions* for locally generating the sounds.

In teleconferencing, of course, 80 percent or more of the video content consists of pictures of people, and people, as yet, cannot be generated at the receiving end. The need is clear, video compression in some form is required, and some of the systems now in use can vastly reduce the amount of information that has to be transmitted. Whereas transmission at regular broadcast standards requires the flow of something like 60 million bits per second (60 Mbs), current compression techniques can reduce this to 1.5 million bits per second (1.5 Mbs). This makes a tremendous difference in the bandwidth required for transmission. And a picture of a person talking looks little different from standard video. A subject in fast motion does break up into a number of successive positions, but these fade quickly into one another. It is certainly useful for teleconferencing purposes, and probably will be applicable also to electronic news gathering and similar purposes, for which the finest possible picture quality and realism is not required.

Most significant of all, the channel bandwidth required for transmission is greatly reduced. For satellite transmission, up to 24 channels of compressed video can be carried on a single transponder, normally intended for one TV channel only. For terrestrial transmission similar substantial savings are possible. For example,

1.5 Mbs video can be transmitted over a T1 telephone line. This is a standard relatively inexpensive line available from AT&T as well as most other carriers throughout the country, commonly in use for data transmission purposes.

The codecs being developed for video compression purposes are in many cases also capable of integrating the audio signal that must accompany the video, as well as some low-speed data transmission, which could be useful as a "back channel" for control purposes (see Chapter 1).

So much for developments that will make teletransmission more practical. Other developments promise improvements in the effectiveness of interactive communication. An imminent development in interactive media is the further integration of computers into systems for the management of conferences. Misar Industries has already incorporated computers into the operation of video meeting room systems. What today is called "computer conferencing" may be known as "alphameric conferencing" tomorrow, since the computer will play a role in the management of conferences using any media, not merely interactive typing.

Computers have already been used in interactive audio conferencing to collect requests to speak, deciding on the basis of programmed criteria who will have the next word, and then simultaneously activating a microphone and displaying the next speaker's nameplate at all meeting locations.

Other media than the alphameric will be given the additional dimension of nonsynchrony. Since all media will probably be transmitted and thus recorded in digital form, eventually they will all be amenable to internal random access. Automatic transfer from voice to print will make it possible to search an audio- or videotape for reference to various key words on the sound track. Eventually we will have automatic pattern recognition — not only optical print scanners such as we have now on the market, but even image recognition, so that the occurrence of specific images may be located and automatically indexed. It could be possible, for example, to set up a camera at a street intersection or airport runway and automatically count the numbers of specific make or model cars or airplanes that pass. Somewhere far off down the road it may be possible to automatically recognize a person's features, and thus to locate the frames in a digital tape in which a particular individual appears.

Of the five major senses, sight and hearing are the ones used in communication and hence are considered the most important. Smell has been used experimentally with motion pictures without marked success. Touch has been used for a type of military radio communication in which members of a combat team wear belts that can convey limited numbers of signals by pressure pulses in various areas. Communication by touch takes place mainly in personal, affectionate, and intimate relationships, except for one custom that is common to most cultures and used both in greeting and social introduction: the handshake. The handshake is generally considered so important in social intercourse that the inability of teleconference participants to use it is often considered a serious and permanent limitation to teleconferencing. But is it really a permanent limitation? Some, who ought to know, say that all the necessary technology for the telehandshake is in existence now; all that is needed is someone to pay the cost of development. We won't predict who will do this, or when it will be done, but neither will we say it will not come to pass. By that time, large-screen three-dimensional TV will already be in use for teleconferencing. A teleconference group will interact with life-size virtual images of the distant participants as though they were real people across the conference table. Visually, and audially, the teleconferencing medium will be all but transparent. The illusion that the distant persons are present will be almost complete.

Take another look at Figure 29 (Chapter 10). These two people are virtually within six feet of each other, so naturally their hands can't seem to meet. But let them approach each other more closely. When people shake hands they approach to within reach of each other, extend a hand and look each other in the eye. Their hands meet and they both "press the flesh," as President Johnson used to say. The hands meet at a point in space that is within a rather limited vertical and horizontal range. Within this window, in the plane of the teleconference screen, the designers will place an artificial hand. It will be warm and will respond according to the exact movements sensed and transmitted by another such hand at the distant conference location. The spot for such introductions or greetings will be to the side of the conference table, where you can approach to within reach of the screen, or whatever plane carries the image. The lasers that produce *your* image are hidden behind the image plane. The artificial hand will not be seen at either end of the line until both of you extend your hands in the usual fashion. Where your hand touches the image hand, the artificial hand will automatically appear in perfect register, and you will seem to be shaking hands with the person in front of you. If he or she produces a weak or limp handshake, that is what you will feel. If the person habitually squeezes enough to make people want to yelp, you will get this treatment here. And the artificial hand will change size, too, according to whether the respondent on the screen has a large, fat, muscular, or thin hand, or whether it is the hand of a female or a child. Science fiction? Possibly, though anyone who has experienced the highly realistic patient simulators now used in medical training will probably agree that this is only a further step in an already amazing technology.

There are other questions, too. Can one person shake hands with many in many locations? Does this have implications for political campaigning? What values will there be in the touch telephone, aside from communication between those who are without both of the primary senses? And what will be the content of private phone calls, after holding hands by phone has become routine?

This may seem like blue-sky speculation. It must be considered, however, that almost all of the technology required is already here and needs only to be adapted and formulated to the purpose. It is almost a truism that today's science fiction is tomorrow's science and technology. With each passing year, moreover, the time gap between science fiction and operating technology grows smaller. This gap may be smaller than we think.

GLOSSARY

Acoustic coupler: A device that generates audial tones to represent a data or facsimile signal fed into it from a keyboard, computer, fax machine, or other source. An ordinary telephone handset is laid on a cradle, receives these tones, and transmits them via a phone line. Another coupler receives and transduces the transmitted information into the form in which it is needed.

Acoustical feedback: Squeal or howl due to sounds from a loudspeaker entering a microphone that feeds that speaker.

Add-on conference feature: Bell telephone service for phones with more than one line. A user can put a conversant on "hold," call another party on a second line, and with a special button retrieve the first caller so that there are three on the line.

Alphameric: Consisting of letters and numerals (plus other symbols and punctuation marks). Same as alpha*nu*meric.

Amplifier: A device that increases the strength of a transmitted signal by transferring the signal modulations from a weak carrier current to a stronger one.

Apparent distance: The distance that a subject seems to be from the viewer, regardless of the size of its retinal image. Compare virtual distance.

ARPA: Advanced Research Projects Agency.

ASCII: American Standard Code for Information Interchange.

Asynchronous: Nonreal time. Two activities, such as the sending and the receiving of a message, do not occur simultaneously.

AT&T: American Telephone and Telegraph Company. Same as Bell Telephone.

Attenuation: Weakening, reduction, thinning out.

Audio: Of or relating to the recording or transmission of sound.

Broad band: Wide bandwidth relative to that required for other media; for example, television and video telephone.

Broadcasting: Transmission over the air of programming for as wide a reception as possible, usually in the home. Compare narrowcasting, closed circuit.

Bandwidth: The range of frequencies in the electromagnetic spectrum allocated or required for a transmission (or recording) channel. Numerically, the difference between the upper and lower frequency bounds of a channel.

Cable TV: A wire television transmission system intended primarily to serve private homes.

Cartridge: (1) A container within which is one reel or roll of film, tape, etc., which is (a) played, received on some external take-up reel, and then rewound, or (b) spliced head-to-tail to make a continuous loop. Compare cassette. (2) Sometimes used as a synonym for cassette.

Cassette: Container within which there are two reels or rolls of tape, film, or the like, a supply reel, and a take-up reel.

CATV: Cable television (formerly community antenna TV).

CB: Citizen band.

Channel: A path or means for one-way transmission; sometimes called line or link. Compare circuit, link.

Citizen band radio: A set of radio frequencies, and equipment for sending and receiving on these frequencies, for the interactive use of private citizens without restriction except transmitter power. Generally used for conversation and informal exchange of information, especially during highway travel.

Closed circuit: A private transmission system totally confined to wire or microwave or the like, not available for reception on standard receiving equipment off-the-air.

Coaxial cable: Two-conductor cable in which one conductor is in the form of a tube or sheath surrounding the other. The two are separated by a dialectric (insulating) material. Used mainly for video transmission.

Common carrier: A company that operates a transmission service without any responsibility for program content (such as telephone companies).

Computer mail: See electronic mail.

Computer teleconferencing: Synchronous or asynchronous alphameric communication using common-access computer files and/or computer-managed interconnections.

Data response: The feedback channel of a two-way system; carries only digital data (pulses) generated by a small number of push buttons or a key pad.

Digital signal: A discrete or discontinuous signal. Amplitude, for instance, is represented by a number or digit, at appropriate time intervals. Compare to analog signal, which is continuous. For example, a digital clock represents time in digits every minute, second, or less, while an analog clock represents the same information by the continuous motion of one or more pointers on a dial.

DOC: Department of Communication (Canadian).

Downstream: From the head end of a CATV system toward the subscribing homes.

Electronic Blackboard: A device developed by the Bell system to sense and transmit chalk markings on a normal-size blackboard of special construction. Display is via TV.

Electronic mail: Typed messages to and from computer users temporarily stored in jointly accessed computer files. Usually programmed to be announced to addressees when they log in and identify themselves.

Facsimile: A medium of transmission for still-visual materials.

FCC: Federal Communications Commission. This body regulates interstate communications. It consists of seven members appointed by the president.

Feedback: (1) See acoustical feedback. (2) Response.

Fiber optics: Thin glass fiber bundles that transmit light rather than current, forming a carrier for TV signals, which modulate the light beam. Fiber optic cable TV systems are capable of far greater length between amplifiers and far greater channel capacity than those based on coaxial cable.

Frame-grabber: An electronic device designed to "reach into" a data stream, and identify and extract only those frames that are addressed to it, or to which it is directed.

Frequency: (1) A measure of cycles per second (Hz) used to divide the continuous electromagnetic spectrum into channels. (2) A channel (band of frequencies) specified by naming the upper and lower boundaries of the band.

Geostationary satellite: A satellite located over the equator, at a height of 22,300 miles, that orbits the earth in a 24-hour period, thus stationary in relation to the surface of the earth.

Handset: The familiar telephone instrument consisting of microphone and speaker and held, usually, in one hand or a shoulder cradle so both hands are free.

Hardware: Equipment.

Head end: Electronic control center and central point of distribution for all or part of a cable TV system.

Headset: A telephone instrument that is worn on the head, consisting of one or two earphones and a small microphone (or acoustical tube leading to a microphone) within an inch or two of the mouth.

Home video: Video playback or playback/record cassette equipment, and/or video disc players, for the home.

HVR: Home video recorder.

Hz: Hertz. One Hertz is equal to one cycle per second.

IDR: Information Dissemination and Retrieval (a Reuters system).

IFTF: The Institute for the Future.

Interaction: Exchange of messages between two (or more) communicants in which each responds to messages from the other.

KC: Kilocycle. Same as kilohertz (KHz) 1000 Hertz.

KHz: Kilohertz.

LDX: A Xerox copier that receives and prints out pages that have been transmitted via facsimile.

Mass medium: A medium through which a single message may reach a large percentage of the people in a given area.

Means of expression: The basic actions that we take in order to communicate — regardless of the presence or absence of a medium to transmit or record these actions or their results.

Message: The content of an intentional communication.

Microwave relay: (1) A point-to-point beamed transmission consisting of one or more links. (2) A system with two or more links. When this definition is used, the word "relay" is not applied to systems with only one link.

Minor media: Media not usually thought of as mass media — less expensive and less glamorous, or usable for more specialized purposes, such as still pictures with sound or alphamerics.

Modulator: A device for the purpose of superimposing the wave form of a TV or other signal onto a carrier wave of much higher frequency (which is either being broadcast or transmitted in a wire, optical fiber, microwave beam, or the like).

MPU: Microprocessor unit. A small computer used in intelligent terminals and home video games.

Multiplexing: (1) Generally, manifold, multiple use. (2) Specifically, sending several messages, in the same or different media, on a single channel, either simultaneously or alternately.

Narrow band: A transmission channel with a bandwidth usually less than one megahertz. Suitable for any media except TV and video telephone.

Narrowcasting: Transmitting programs intended for specific relatively small audiences, such as physicians, students enrolled in a particular course, the financial community, or the like.

NASA: National Aeronautics and Space Administration.

Off-the-air: Received from a broadcast on a standard TV or radio channel.

Open audio: A system in which all microphones are on — anyone may interrupt and be heard at any time.

PAL: Phase alteration line. A system of color TV developed in Europe.

Parallactic angle: An angle measuring the difference in direction between camera and monitor from an interactive video user.

Pay TV: Broadcast or cablecast service that can only be received by paying a special fee, either by the channel, by the day, or by the program.

PBS: Public Broadcasting Service.

Picturephone: A desk-top unit and transmission service field tested by Bell Telephone. The terminal unit consisted of camera, monitor, and controls. Earlier models incorporated an audio system.

Picturephone Meeting Service: Telephone conference service offered by Bell Telephone since 1976.

PL: Private line.

Plasma panel: Computer terminal display device capable of 500-line (TV) definition but not so far capable of representing gray scale or color.

Plato: A computer-managed instructional system incorporating plasma panel display and interactive software.

POTS: Plain old telephone service.

Private line: Audio channel linking various points of origination for technical and operational coordination.

Program: A planned and usually scheduled unit of presentation, usually of a one-way nature, for media or face-to-face display.

Proxemics: The use of space by human beings; the study of distance between persons in terms of their intercommunication.

PTT, PT&T: Public telephone and telegraph.

QUBE: A thirty-channel CATV service operated by Warner Amex Cable Corp in several U.S. cities (QUBE is not an acronym; the name is arbitrary).

QWERTY: Standard typewriter keyboard, named for the first six letters on the top row.

Real time: (1) A situation in which two or more participants interact at the same time—when one "sends" the others "receive," and so on. (2) A situation in which two or more participants interact at the same time — all may send at once, all receive at once, or any combination in between.

RMT: Remote Meeting Table.

Semaphore: One of various systems of communicating visually at distances beyond earshot and not beyond line of sight.

Semimotion: The motions of writing, drawing, printing, pointing, flashing, and the like.

Social presence: The degree to which a distant interactant seems to be present during a teleconference.

Software: Any component of a medium that is not equipment (hardware): programs, recordings, transmitted messages, films, tapes, and so on.

Speakerphone: A desk-top telephone unit that can be used at a distance of several feet, consisting of a nondirectional microphone and a small speaker.

Split screen: Pictures from two or more sources displayed in different areas of the same screen.

Spring loaded: Containing a spring that returns a control to its normal position, for example, a push button that remains "on" only as long as it is pushed.

Synchronous: Taking place simultaneously.

Tektronix TV recorder: A hard-copy recorder produced by the Techtronics Corp. (model 4632) for recording a still image, on paper, of a television signal.

Telautograph: A device to transmit and/or receive writing, drawing, printing, and so on.

Telecommunication: (1) Live media communication. (2) Communication via electronic media whether live, delayed, or recorded and played back.

Teleconference: A conference among three or more intercommunicating persons, at least one of whom is separated from the others by space and/or time and/or other barriers.

Telegroup: A number of isolated individuals who are joined by teleconferencing.

Telephone bridge: A device that allows a telephone operator to join up to thirty participants in an open audio connection.

Teletext: Various systems of transmitting text on the same channel with a TV broadcast. The entire data base of text (and simple graphics) is transmitted and constantly recycled; users select the pages they want for display.

Telewriting: See telautograph.

Thermofax: A copy machine that uses special paper and a heat process.

Touch-tone pad: A telephone address selection device, consisting of 12 buttons, which generates tones of different frequencies to represent the numbers, letters, or other symbols transmitted.

Touch-tone teletext: A system for selection (by phone) and transmission (by teletext) of text pages on demand. Each page so selected is transmitted only once, not recycled.

Upstream: Flow of communication away from subscriber homes in a cable TV system and toward the head end.

VCR: Video cassette recorder.

Video: (1) Technically, the picture portion of the television signal. (2) Synonym for television, both picture and sound.

Video cassette: A container holding two reels or rolls (supply and take-up) of videotape. Compare cartridge.

Video cassette recorder: A machine to record (and usually also to play back) video cassettes.

Videodisc: A recording of TV sound and picture on a disc similar to a phonograph record.

Videotape: Mylar or similar tape coated with a magnetic oxide in widths of two inches (old broadcast standard) to one-half inch (home VCR standard).

Videotex: (1) Various systems of information retrieval for the home, fed via the telephone and displayed, like teletext, on the home TV screen. Only the pages users select are fed to their homes. Compare teletext. (2) In Britain, both systems described in definition 1 and teletext are covered by the term.

Viewdata: A British term for systems that transmit text and graphic information to the home via telephone wire. Same as videotex.

Virtual distance (of an image): The distance from a viewer that a real subject would have to be to produce a retinal image the same size. Compare apparent distance. (Virtual: being in essence or effect, but not in fact.)

VTR: Videotape recorder.

REFERENCES

ANKER-JOHNSON, B. (1974) Telecommunication — A Key to Resolving Our Energy Shortage. Washington, DC: Department of Commerce, Office of Telecommunications.

BAKER, E. (1978) "Interactive video teleconferencing systems: Will they gain acceptance or disappear?" Presented at the meeting of the American Psychological Association.

Bell System (n.d.) Phonepower. New York: Author. (leaflet)

BRETZ, R. [ed.] (1976) A Teleconference on Teleconferencing. Working paper ISI/WP-4. Los Angeles: University of Southern California, Information Sciences Institute.

——— and L. DOUGHERTY (1974) Two-Way TV Teleconferencing for Government: The MRC-TV System. R1489-MRC. Santa Monica, CA: Rand Corporation.

BROWN, H. G. et al. (1978) A General Description of Telidon: A Canadian Proposal for Videotex Systems. Note 697E. Ottowa: Canadian Department of Communications.

CHAPANIS, A. (1975) "Interactive human communication." Scientific American 232 (March).

CHRISTIE, B. and M. ELTON (1975) Research on the Differences Between Telecommunication and Face-to-Face Communication in Business and Government. London: Communications Study Group.

DICKSON, E. M. and R. BOWERS (1973) The Video Telephone. Washington, DC: National Science Foundation.

DOLMATCH, T. B. [ed.] (1979) Information Please Almanac, 1979. New York: Information Please Publishing.

Electronic Conventions, Inc. (1979) Interactive Telecommunications Systems in Social Uses. El Segundo, CA: Author.

ELTON, C. J. and J. CAREY (1979) Implementing Interactive Telecommunication Services. New York: Alternate Media Center.

FLESCH, R. (1949) The Art of Readable Writing. New York: Colliers.

GOLDSMITH, A. (1974) Telecommunications — An Alternative to Travel. Washington, DC: U.S. Department of Commerce, Office of Telecommunications.

——— (1969) Telecommunications for Advanced Metropolitan Function and Form. Washington, DC: National Academy of Engineering, Committee on Telecommunications.

HALL, E. T. (1966) The Hidden Dimension. Garden City, NY: Doubleday.

HARKNESS, R. C. (1973) Telecommunications Substitutes for Travel. Washington, DC; U.S. Department of Commerce, Office of Telecommunications.

HOUGH, R. W. and R. R. PANKO (1977) Teleconferencing Systems: A State of the Art Survey and Preliminary Analysis. Menlo Park, CA: Stanford Research Institute.

Intermedia (1979) Publication of the International Institute of Communications. Vol. 7 (May).

JOHANSEN, R., M. McNULTY, and B. McNEAL (1978) Electronic Education: Using Teleconferencing in Postsecondary Organizations. Report R2. Menlo Park, CA: Institute for the Future.

Joint Unit for Planning Research (1973) Communications Studies Group Final Report. London: University College.

LUCAS, W. et al. (1979) The Spartanburg Interactive Cable Experiments in Home Education. R2271. Santa Monica, CA: Rand Corporation.

McLUHAN, M. (1964) Understanding Media. New York: McGraw-Hill.

MARTIN, J. (1971) Future Developments in Communications. Englewood Cliffs, NJ: Prentice-Hall.

MONSON, M. (1978) Bridging the Distance. Madison: University of Wisconsin Extension.

——— M. PARKER, and M. RICCOMINI [eds.] (1977) A Design for Interactive Audio. Madison: University of Wisconsin.

Online Review, the International Journal of Online and Teletext Information Systems (1979) Vol. 3 (January).

PARK, B. (1974) An Introduction to Telemedicine: Interactive Television for Delivery of Health Services. New York: Alternate Media Center.

ROBERTS, M. et al. (monthly) The Videocassette and CATV Newsletter. Beverly Hills, CA: Author.

SHORT, J., E. WILLIAMS, and B. CHRISTIE (1976) The Social Psychology of Telecommunication. New York: John Wiley.

STAPLEY, B. (1972) "Visual enhancement of telephone conversations." Ph.D. dissertation, Imperial College, London.

University of Wisconsin Extension (n.d.) Teletechniques: Instructional Design for Teleconferencing, a Handbook. Madison: Author.

——— Center for Interactive Instructional Programs (n.d.) Technical Design for Audio Teleconferencing. Madison: Author.

University of Wisconsin Extension, Division of Educational Communications (n.d.) The Status of the Telephone in Education. Report on the Second Annual International Communications Conference. Madison: Author.

——— (n.d.) The Telephone in Education, Book II. Report on the Third Annual International Communications Conference. Madison: Author.

VALEE, J. et al. (1974a) Group Communication Through Computers, Vol. 1. Menlo Park, CA: Institute for the Future.

——— (1974b) Group Communication Through Computers, Vol. 2. Menlo Park, CA: Institute for the Future.

INDEX

Acoustic coupler, 97-98
Acoustical feedback, 72-77; diagram, 75
Alpha-geometric graphics, illustrated, 151
Alpha-mosaic graphics, illustrated, 150-151
Alphameric media systems, 93-109
ANTIOPE, 149
Apparent distance (apparent proximity), 184-186, 219
Asynchronous systems, 101-109; international schedules, 106
Audio systems versus video, 232-239; table: comparative satisfaction, 233
Australian Confravision, 54-56; multiplexing system diagram, 55

Back channel (in Reading system), 64-65, 68
Backup channels, 233, 240-241
Bankers Trust video system: an example of proxemics, 191-197; illustrated, 193
Bathtub curve, 223-225; illustrated, 224
BBC Teletext services, 143-148
Bell Canada video system, 232; table: comparative satisfaction, 233
Bell Laboratories experimental video conferencing system, 38-42
Bell System Electronic Blackboard, 88-92
Boeing audiovisual systems, 84-86
Brain hemispheres, 15, 139

CAI, CMI systems, 139
Carlstedt, James, 100
Carterfone decision, 72
CEEFAX, 143-148, 156, 163
Channels required for unswitched system, 36; diagrammed, 37
Chapanis, Alphonse, 109
Closed user group (application of PRESTEL), 160; diagrammed, 158-159
Codec, 242
Colorado Video, 126, 129, 133
Communication defined, 19
Communication Study Group: See CSG
Communication theory, 18
Computer conferencing: compared with correspondence, 102-104; compared with face-to-face meetings, 105; compared with spoken language, 104;

anonymous entry, 107; private entry, 107-108
Computer-managed teleconferencing, 244
Conference activities (face-to-face), 236-238
Confravision, 33-37, 232; photo, 35; an example of proxemics, 187-191; compared to Picturephone, 194-195; eye contact comparison table, 204; table: comparative satisfaction, 233
Control key on keyboard, 96
Convenience (a factor in user acceptance), 221-222
Correspondence as interaction, 21
Cover shot, 214-215
CSG (Communication Study Group) studies, 185, 196, 231-239
Cutting problems, 215-218; illustrated, 216

Darome Convener, 74-76, 119
Data response, 110-116; FCC requirements, 115
Delta component, 242-243; defined, 242

Edex (classroom response system), 111
Electronic Blackboard, 88-92; photo, 90
ELIZA, 15
Engineering approach to communication, 18
Establishing shot, 215
Extrinsic factors, 167
Eye contact, 197-208

Face-to-face conferences at various proximities, 196-197; illustrated, 198
Face-to-face simulated, 176-178
Facsimile and slow scan TV, 120; use in NASA and Boeing systems, 84
FCC data response requirements, 115
Flesch, Rudolf, 107
Fog Index, 107; defined, 109
Forum (computer conferencing system), 101-104; conferences analyzed, 102, 103
Frame grabber, 147
Freeze-frame video, 118

Genesys, 57

Hall, Edward T., 182, 196
The Hidden Dimension, 182
Hughes Aircraft Co. slow scan system, 119
Humanizing, 170

IBA teletext services, 143-146
IDR (Reuters), 149-153
Information provider, defined, 156; diagrammed, 158-159
Institute for the Future, computer conference studies, 102-106; table summary, 102
Interaction, defined, 12-13
Intimate distance, defined, 182-183; chart, 188-189
Intrinsic factors, 167, 174-219

Keyboards, 95-98
KSL-TV touch-tone teletext, 164

Legibility of print on slow scan TV, comparative photos, 130-131
Link mode of interaction, 98

Mail correspondence as interaction, 21

Maintenance (a factor in operational support), 223-225; illustrated, 224
Mass media named, 10
Means (of communication), 169-170; distinguished from media, 23-24, 25; nonverbal, 28
Media (of communication) distinguished from means, 22-24, 25; learning to use a one-way medium, 27
Media skills, 26-27
Media system model, 25
Medium, defined, 19
Meeting tasks, comparative table, 237
Message, defined, 16-18; message style, 170
Misar II system, 65-68; elevation drawing, 66; plan of typical meeting room, 67; simulated handshake, 200-201; illustrated, 201; eye contact comparison table, 204
Monitors, 208-213
Monson, Mavis, 170-171
Mosaic characters, 148; illustrated, 150-151
MRC-TV video system, 56-60, 191; eye contact comparison table, 204
Multiplexing of video signals, 241

National Academy of Engineering substitutability study, 226-227
NASA audiovisual system, 84-86
Newstime cable TV channel, 118, 125, 129
Nippon Telephone and Telegraph experimental video system, 56
Nonhuman intelligence, 14-15
Nonredundant transmission, 136
Nonverbal communication means, 28

Operational support, 222-225
ORACLE, 143-146, 156
Orientation shot, 215
Overall shot, 213-215

Participation, 170
Party line (interactive system), 99-100
PDIs (Picture description instructions), 163; illustrated, 165
Personal distance: defined, 182-183; chart, 188-189; illustrated, 190
Picturephone, 40-42; illustrated, 41
Picturephone Meeting Service, 45-54; overview shots, 46; Los Angeles facility general photo, 48; easel camera (Photo), 49; graphics stand and camera (photo), 50; graphic stand table (photo), 51; control panel (photo), 52; recording equipment (photo), 53; eye contact comparison table, 204; monitor use, 213-214
Picture Meeting Service, first generation, 42-44; diagram, 43; distinguished from Picturephone Meeting Service, 68
Picture quality and frame frequency (table), 127
Pointing, on slow scan pictures, 132
POTS (plain old telephone service), 69-70
PRESTEL, 154-161
Preventive maintenance, 222-223

Printed materials in teleconferencing, 83-84
Program, defined, 23
Projected materials in teleconferencing, 84-85
Prompters (script), 206
Proxemics, 181-197; and visual media, 184-187; applied to Confravision, 187-191; applied to Bankers Trust system, 191-197
PT&T teleconference systems, 30
Public distance: defined, 182-183; chart, 188-189; illustrated, 190

Quasi-interactive media systems, 137-166
QUBE data response system, 110-114

Radio, compared with telephone, 11
RCA Global Communication, Video Voice system, 122
Reading, PA, video system, 60-65; eye contact comparison table, 204; monitor use, 210
Real time interaction, 98-101
Recording (storage) media, use in interaction, 21
Remote Meeting Table: See RMT system
Responsive communication, 17
Reuters IDR, 149-153
RMT system, 81-83, 232; table: comparative satisfaction, 233
Robot Research, 126-127, 129
Row grabber, 150

St-Hubert, Quebec, cable TV system, 114, 140
Satellite distribution, 241
Screen widths (measure of viewing distance), 184, 218
Selected response systems: QUBE, 110-114; Spartanburg, NC, system, 111-112
Semimotion, defined, 87
Shift key on keyboard, 96-97
Shot content, 184
Simulated interaction, 14
Simulation of face-to-face communication, 176-178
Slow scan TV, 117-136; how the system works, 122-124
Social distance: defined, 183-184; chart, 188-189; illustrated, 190
Social presence, 121
Spartanburg, NC, project, 111
Speaker identification, 80-81
Speakerphones, 71-72
Split screen, in MRC-TV system, 58
Spoken word, 24; as basis for classification of media, 29-30
Stapley, Barry, 185, 196, 219
Still content of motion pictures, 120-121
Still images as a medium, 120-122
Still/motion continuum, 135-136
Storage and retrieval systems, 139-141
Storyteller, 14
Style of expression continuum, from informal conversation to edited composition, 106-107

Talkomatic, 99
Teaching machines, 12, 137-139
Telautograph, 87
Teleconference, defined, 20
Telehandshake, 245-246
Telemedicine, as substitute for travel, 228
Telephone: compared to radio, 10-12; WATS lines, 21; conference calls, 70-72

Teletext, 143-149; graphics, 148; illustrated, 150-151; diagram, 158; touch-tone teletext, 164
Teletype, telex, 94-95; interactive uses, 95
Telewriting, 87-88, 94
Telidon, 161-163
3-D video conferencing, 199-200
TITAN, 149, 161
Touch communication, 245
Touch-tone teletext, 164, 166
Translation, automated, 180
Transparency, of a medium, 27-28
Travel substitutability, 225-230
Typing skills, 95-96

Umbrella provider, 158; diagrammed, 159
University of Quebec audio system, 232; table: comparative satisfaction, 233
University of Wisconsin Educational Network (ETN), 72-79
Unswitched systems, 36
User acceptance, 219-222

Video compression, 242-243; in slow scan TV, 126
Video conferencing system, coaxial (proposed) illustrated: individual mode, 207; group mode, 209
Videophone, first services, 230
Video versus audio systems, 232-239; table: comparative satisfaction, 233
Videotex, 154-166; diagrammed, 158-159
Video Voice, 122
Viewdata, distinguished from videotex, 141
Viewing distance and apparent distance, 184, 187, 219; chart, 188-189; illustrated, 186, 190
Virtual distance (virtual proximity): defined, 185; diagrammed, 186; chart, 188-189; investigated, 219; illustrated, 190
Voice-controlled TV tuning and computer input, 180-181

Western Electric 50A Speakerphone, 72
Whisper mode of computer conference entry, 107-108
Winkie, 14
Wipe effect in slow scan TV, 125
Women's cable TV channel, 119